100 PLANTS FOR BEEKEEPERS

Compiled by Stuart A. Roberts

100 Plants for Beekeepers
© Stuart Roberts

All rights reserved. No part of this publication may be reproduced, stored in a retrieval system, transmitted in any form or by any means electronic, mechanical, including photocopying, recording or otherwise without prior consent of the copyright holders.

ISBN:
Soft back: 978-1-914934-38-4
Hard back: 978-1-914934-39-1

Published by Northern Bee Books, 2022
Scout Bottom Farm
Mytholmroyd
Hebden Bridge
HX7 5JS (UK)

Original book design by Zoe McCullagh-George
Artwork by DM Design and Print

All of the photographs were taken by Stuart Roberts. It is a UK centric book written with my own experiences in mind. So the flowering times relate to the UK Midlands and will, therefore, differ to the south coast and the north of Scotland.

ACKNOWLEDGEMENTS

In loving memory of Geoff & Mary Hopkinson for all the help and support they have given me.

Contents

Glossary	v
Flowering Plant Biology	1
Attracting Pollinators	3
Plant Families	6
Flowering Periods	17
Plants by Main Flowering Periods	22
Spring	23
Summer	65
Autumn	112
Winter	116
Plant Index	127

Glossary

Annual: An annual plant is a plant that completes its life cycle, from germination to the production of seeds, within one growing season, and then dies.

Axil: The upper angle between one part of a plant and another, e.g. the stem and a leaf.

Axillary: Borne in or arising from the axil of a leaf.

Basal: Situated or attached at the base.

Biennial: A plant which completes its life cycle (germinates, reproduces and dies) within two years; usually also forms a basal rosette of leaves the first year and flowers and fruits the second year.

Blade: The lamina or flattened part of a leaf, excluding the stalk.

Bracts: A modified leaf associated with a flower or inflorescence and differing in shape, size or colour from other leaves (and without an axillary bud).

Calyx (pl. calyces): A collective term for the sepals of one flower; the outer whorl of a flower, usually green. Compare corolla.

Capitulum (pl. capitula): A compact head of a structure, in particular a dense flat cluster of small flowers or florets, as in plants of the daisy family.

Carpel: The female reproductive organ of a flower, consisting of an ovary, a stigma, and usually a style. It may occur singly or as one of a group.

Catkin: A catkin is a slim, cylindrical flower cluster, with inconspicuous or no petals.

Cordate: Heart-shaped, with the notch lowermost; of the base of a leaf.

Corolla: A collective term for the petals of a flower. Compare calyx.

Corona: In flowering plants, a ring of structures that may be united in a tube, arising from the corolla or perianth of a flower and standing between the perianth lobes and the stamens. The trumpet of a daffodil is a corona.

Corymb: An inflorescence with branches arising at different points but reaching about the same height, giving the flower cluster a flat-topped appearance. See 'Wild Carrot'

Cultivar: A term derived from "cultivated variety" denoting an assemblage of cultivated plants clearly distinguished by one or more characters

Cyme: A type of inflorescence in which the main axis and all lateral branches end in a flower

Decussate: Each leaf pair is produced at right angles to the next

Dentate: Toothed, especially in reference to leaf margins.

Dicotyledon (adj. dicotyledonous): A flowering plant whose embryo has two (rarely more) seed leaves.

Dioecious: Of vascular plants, when male and female reproductive structures develop on different individual plants

Ensiform: Shaped like the blade of a sword.

Exine: The decay-resistant outer coating of a pollen grain or spore.

Fascicle: A cluster, e.g. a tuft of leaves all arising from the same node.

Genus (pl. genera): A group of one or more species with features, ancestry or both in common. Genus is the principal category of taxa intermediate in rank between family and species.

Glabrous Lacking surface ornamentation such as hairs, scales or bristles; smooth

Globose: Approximately spherical.

Herbaceous: Not woody; usually green and soft in texture.

Inflorescence: Several flowers closely grouped together to form an efficient structured unit; the grouping or arrangement of flowers on a plant.

Involucre (adj. involucral): A structure surrounding or supporting, usually, a head of flowers.

Lanceolate: Longer than broad, narrowly ovate. Broadest in the lower half and tapering to the tip, like a lance or spear head.

Laticiferous: Latex-bearing, producing a milky juice.

Leaflet: The ultimate segments of a compound leaf.

Ligulate: Strap-shaped.

Lobe: Part of a leaf, often rounded, formed by incisions to about halfway to the midrib.

Margin: The edge, as in the edge of a leaf blade.

Midrib: The central and usually most prominent vein of a leaf or leaf-like organ.

Monocot: An abbreviation of monocotyledon

Monocotyledon: A flowering plant whose embryo contains one cotyledon (seed-leaf)

Oblanceolate: Having a lanceolate shape but broadest in the upper third.

Ovate: Shaped like a section through the long-axis of an egg, attached by the wider end.

Palmate: Leaf with veins radiating out from a central point, resembling spread out fingers pointing away from the palm.

Panicle: A compound raceme; an indeterminate inflorescence in which the flowers are borne on branches of the main axis or on further branches of these.

Pedicel: The stalk of a flower; may also be applied to the stalk of a capitulum in the Asteraceae.

Pendent: Hanging down or overhanging

Pentamerous: In five parts; particularly with respect to flowers, five parts in each whorl.

Perennial: A plant whose life span extends over several years.

Perfect: Of a flower, when bisexual.

Petiole: The stalk of a leaf.

Phyllary: An individual bract within an involucre or involucel.

Pinnate: A compound leaf with leaflets arranged on each side of a common petiole or axis; also applied to how the lateral veins are arranged in relation to the main vein.

Pistillate: Describing a flower containing pistils but no stamens.

Raceme: An indeterminate inflorescence in which the main axis produces a series of flowers on lateral stalks, the oldest at the base and the youngest at the top. Compare spike.

Reticulate: An indeterminate inflorescence in which the main axis produces a series of flowers on lateral stalks, the oldest at the base and the youngest at the top. Compare spike.

Rhizome (adj. rhizomatous): A perennial underground stem usually growing horizontally.

Rugose: Wrinkled.

Salverform: Trumpet-shaped; having a long, slender tube and a flat, abruptly expanded limb.

Scale: A reduced or rudimentary leaf, for example around a dormant bud.

Scape: A stem-like flowering stalk of a plant with radical leaves.

Sepals: In a flower, one of the segments or divisions of the outer whorl of non-fertile parts surrounding the fertile organs; usually green. Compare petal.

Septum: A partition, e.g. the membranous wall separating the two valves of a pod.

Serrate: Toothed with asymmetrical teeth pointing forward; like the cutting edge of a saw.

Sessile: Attached without a stalk, e.g. of a leaf without a petiole or a stigma, when the style is absent.

Sheath: A tubular or rolled part of an organ, e.g. lower part of the leaf in most grasses.

Shrub: A woody perennial plant without a single main trunk, branching freely, and smaller than a tree.

Spp.: Abbreviation of 'species'.

Stamen (adj. staminate): The male organ of a flower, consisting (usually) of a stalk called the filament and a pollen-bearing head called the anther.

Stipule: A small appendage at the bases of leaves in many dicotyledons.

Sub-shrub: A small shrub which may have partially herbaceous stems, but generally a woody plant less than 1 metre high.

Tepal: A perianth segment, either sepal or petal; usually used when all perianth segments are indistinguishable in appearance.

Tree: A woody plant, usually with a single trunk and generally more than 2–3 metres tall.

Trifoliate: A compound leaf of three leaflets; for example, a clover leaf. Sometimes *trifoliolate*.

Umbel: A racemose inflorescence in which all the individual flower stalks arise in a cluster at the top of the peduncle and are of about equal length; in a simple umbel, each stalk is unbranched and bears only one flower. A cymose umbel looks similar to an ordinary umbel but its flowers open centrifugally.

Venation: The arrangement of veins in a leaf.

Verticillate: Arranged in one or more whorls, i.e. several similar parts arranged at the same point of the axis, e.g. leaf arrangement.

Verticillaster: A type of pseudoverticillate inflorescence, typical of the Lamiaceae, in which pseudo-whorls are formed from pairs of opposite cymes.

Whorl: A ring of organs borne at the same level on an axis, for example leaves, bracts or floral parts.

Flowering Plant Biology

Flowering plants or angiosperms emerged some 160 million years ago in the Lower Cretaceous period. They dominate the Earth's surface and are a very important food source for insects, birds and mammals. In addition, they are the most economically important group of green plants, serving as a source of pharmaceutical, fibre products, timber, ornamentals, and other commercial items.

The clever part is that there are two kinds of reproductive cells produced by flowers. Microspores, which will divide to become pollen grains, are the "male" cells and are borne in the stamens (or microsporophylls). The "female" cells are called megaspores, which will divide to become the egg cell (megagametogenesis). These are contained in the ovule and enclosed in the carpel.

While the majority of flowers are perfect or hermaphrodite (having both pollen and ovule producing parts in the same flower structure), flowering plants have developed numerous morphological and physiological mechanisms to reduce or prevent self-fertilization. Heteromorphic flowers have short carpels and long stamens, or vice versa, so animal pollinators cannot easily transfer pollen to the pistil (receptive part of the carpel). Homomorphic flowers may employ a biochemical (physiological) mechanism called self-incompatibility to discriminate between self and non-self pollen grains. In other species, the male and female parts are morphologically separated, developing on different flowers. For fertilisation to occur the pollen grain is deposited on the stigma of the flower. The pollen then germinates and the pollen tube protrudes between the gaps in the hard outer layer (exine) of the pollen and grows down the style to the ovary. The male gamete fertilises the ovum to produce an embryo.

There are a number of ways for the pollen grain to reach the stigma. Pollen from the anther reaching the stigma in the same flower is self-pollination, which leads to more uniform progeny, meaning that the species is, for example, less resistant as a whole to disease. However, it does not need to expend energy on attracting pollinators and can spread beyond areas where suitable pollinators cannot be found.

Alternatively there is cross-pollination. This requires pollen from another flower or plant reaching the stigma, allowing for greater genetic diversity and therefore hybrid vigour. This method requires some mode of transportation of which these are the four most common:

Entomophily – insect pollination   Hydrophily – water pollination
Anemophily – wind pollination   Zoophily – vertebrate pollination

Attracting Pollinators

From around 130 million years ago flowering plants and insects have evolved together in something called co-evolution. This has resulted in a number of different ways of attracting insects to the plants.

Entomophilous plant species have frequently evolved mechanisms to make themselves more appealing to insects, e.g. brightly coloured or scented flowers, nectar, or appealing shapes and patterns. Pollen grains of entomophilous plants are generally larger than the fine pollens or anemophilous (wind-pollinated) plants, which has to be produced in much larger quantities because such a high proportion is wasted. This is energetically costly, but in contrast, entomophilous plants have to bear the energetic costs of producing nectar.

Butterflies and moths have hairy bodies and long proboscides which can probe deep into tubular flowers. Butterflies mostly fly by day and are particularly attracted to pink, mauve and purple flowers. The flowers are often large and scented, and the stamens are so-positioned that pollen is deposited on the insects while they feed on the nectar. Moths are mostly nocturnal and are attracted by night-blooming plants. The flowers of these are often tubular, pale in colour and fragrant only at night. Hawkmoths tend to visit larger flowers and hover as they feed; they transfer pollen by means of the proboscis. Other moths land on the usually smaller flowers, which may be aggregated into flower heads. Their energetic needs are not as great as those of hawkmoths and they are offered smaller quantities of nectar.

Flowers pollinated by bees and wasps vary in shape, colour and size. Yellow or blue plants are often visited, and flowers may have ultra-violet nectar guides, that help the insect to find the nectary. Some flowers, like sage or pea, have lower lips that will only open when sufficiently heavy insects, such as bees, land on them. With the lip depressed, the anthers may bow down to deposit pollen on the insect's back. Other flowers, like tomato, may only liberate their pollen by buzz pollination, a technique in which a bumblebee will cling on to a flower while vibrating its flight muscles, and the dislodges the pollen. Because bees care for their brood, they need to collect more food than to just maintain themselves and, therefore, are important pollinators. Other bees are nectar thieves and bite their way through the corolla in order to raid the nectary, in the process bypassing the reproductive structures.

Some plant species co-evolved with a particular pollinator species, such as the bee orchid. The species is almost exclusively self-pollinating in its northern ranges but is pollinated by the solitary bee Eucera in the Mediterranean area. The plant attracts these insects by producing a scent that mimics the scent of the female bee. In addition, the lip acts as a decoy, as the male bee confuses it with a female that is visiting a pink flower. Pollen transfer occurs during the ensuing pseudo-copulation.

Inflorescences pollinated by beetles tend to be flat with open corollas or small flowers clustered in a head with multiple, projecting anthers that shed pollen readily. The flowers are often green or pale-coloured, and heavily-scented, often with fruity or spicy aromas, but sometimes with odours of decaying organic matter. Some, like the giant water lily, include traps designed to retain the beetles in contact with the reproductive parts for longer periods.

When a honeybee lands on a Birdsfoot Trefoil blossom, it stands on the fused side petals looking for nectar, the bee's feet work the petals open exposing a keel petal inside. When the bee steps down onto the keel, its weight triggers the keel to bounce down – like a trampoline – and the petals spread. Then the stamens inside the keel that are covered with pollen will pop up and dust the bee's belly.

In horse chestnut, the flowers are mostly white with nectar-guides which are yellow at first, turning to a deep crimson once the flower is pollinated. This is significant because the bees cannot see the colour red so at that point the flower is effectively 'switched off' as far as they are concerned.

Other Nectar Guides

Nectar guides can be both visible and invisible to the human eye but more importantly they are very visible to the bees in the ultraviolet spectrum. This adaptation benefits both the flower (more efficient pollination) and the bee (rapid collection of nectar).

Plant Families

Amaryllidaceae

The largest genera of the family are Allium (260-690 species), Nothoscordum (25) and Tulbaghia (22). Characterised by simple or prolific bulbs, sometimes with lateral rhizomes. Leaf sheaths are long, the tepals free and the corona absent.

Species of importance to bees:

Chives, *Alluim schoenoprasum* *96*

Snowdrop, *Galanthus nivalis* *122*

Apiaceae

Apiaceae is a family of mostly aromatic flowering plants named after the type genus Apium and commonly known as the celery, carrot or parsley family. It is the 16th-largest family of flowering plants, with more than 3700 species in 434 genera, including well-known and economically important plants. Most Apiaceae are annual, biennial or perennial herbs, though a minority are woody shrubs or small trees. Their leaves are frequently aggregated toward the base.

Species of importance to bees:

Wild Carrot, *Daucus carota* *85*

Lovage, *Levisticum officinale* *54*

Garden Angelica, *Angelica archangelica* *91*

Aquifoliaceae

Ilex, or holly, is a genus of 400 to 600 species of flowering plants in the family Aquifoliaceae, and the only living genus in that family. The species are evergreen or deciduous trees, shrubs, and climbers from tropics to temperate zones worldwide.

Species of importance to bees:

Holly, *Ilex aquifolium* *47*

Araliaceae

The Araliaceae is a family made of 53 genera and 700 species of flowering plants including perennial herbs, trees, vines and succulents. The family has large, usually alternate leaves, five-petalled flowers arranged in clusters and berries. Some genera are used as foliage plants.

Species of importance to bees:

Ivy, *Hedera helix* *115*

Asparagaceae

Asparagaceae is a family of flowering plants, placed in the order Asparagales of the monocots. Asparagaceae includes 114 genera with a total of around 2900 known species.

Species of importance of bees:

    Bluebell, *Hyacinthoides non-scripta* *32*

    Asparagus, *Asparagus officinalis* *42*

Asteraceae

Asteraceae is a very large and widespread family of flowering plants (Angiospermae). The family currently has 32,913 accepted species in 1,911 genera and 13 subfamilies. In terms of numbers of species, the Asteraceae are rivalled only be the Orchidaceae. Many members have composite flowers in the form of flower heads surrounded by bracts. When viewed from a distance, this may have the appearance of being a single flower.

Species of importance to bees:

    Dandelion, *Taraxacum officinale* *28*

    Ragwort, *Senecio jacobaea* *106*

    Sunflower, *Helianthus spp.* *104*

    Kanpweed, *Centaurea nigra* *103*

    Sea Aster, *Tripolium pannonicum* *107*

    Dhalia, *Dhalia spp.* *89*

    Tansy, *Tanacetum vulgare* *105*

    Golden Rod, *Solidago spp.* *101*

    Rudbeckia, *Rudbeckia fulgida* *92*

    Yarrow, *Achillea millefolium* *93*

Balsaminaceae

The Balsaminaceae (commonly known as the balsam family) are a family of dicotyledonous plants, comprising of two genera: the Impatiens with over 1000 species and the Hydrocera, which features only one species. The flowering plants may be annual or perennial. They are found throughout temperate and tropical regions, primarily in Asia and Africa, but also North America and Europe. Notable members of the family include Jewelweed and Busy Lizzie.

Species of importance to bees:

Himalayan Balsam, *Impatiens glandulifera* *87*

Berberidaceae

The Beriberidaceae are a family of 18 genera of flowering plants commonly called the barberry family. The family is in the order Ranunculales. The family contains about 700 known species, of which the majority are in Berberis. The species include trees, shrubs and perennial herbaceous plants.

Species of importance to bees:

Mahonia, *Mahonia spp.* *117*

Barberry, *Berberis spp.* *31*

Boraginaceae

Boraginaceae, the Borage or Forget-Me-Not family, includes a variety of shrubs, trees and herbs, totalling about 2000 species in 146 genera found worldwide. Most pollination is by hymenopterans, such as bees. Most species have inflorescences that have a coiling shape, at least when new. The flower has a usually five-lobed calyx. The corolla varies in shape from rotate to bell-shaped to tubular, but generally have five lobes. It can be green, white, yellow, orange, pink, purple or blue.

Species of importance to bees:

Borage, *Borago officinalis* *77*

Forget-Me-Not, *Myosotis spp.* *40*

Vipers Bugloss, *Echium vulgare* *63*

Phacelia, *Phacelia tanacetifolia* *61*

Common Lungwort, *Pulmonaria officinalis* *24*

Comfrey, *Symphytum spp.* *53*

Brassicaceae

Brassicaceae, or Cruciferae, is a medium-sized and economically important family of flowering plants commonly known as the mustards, the crucifers, or the cabbage family. Most are herbaceous plants but there are some shrubs. Flowers have four free sepals, four free alternating petals, two short and four longer free stamens. The fruit has seeds in rows, separated by a thin wall (or septum).

Species of importance to bees:

Oilseed Rape, *Brassica napus* *35*

Charlock, *Sinapis arvensis* 57

Aubretia, *Aubretia deltoidea* 26

Sweet Alison, *Lobularia maritima* 88

Caprifoliaceae

The Caprifoliaceae or honeysuckle family are dicotyledonous flowering plants consisting of about 860 species in 42 genera, with a nearly cosmopolitan distribution. Centres of diversity are found in eastern North America and eastern Asia, while they are absent in tropical and southern Africa. The flowering plants in the family are mostly shrubs and vines: rarely herbs. They include some ornamental garden plants grown in temperate regions. The leaves are mostly opposite with no stipules and may be either evergreen or deciduous. The flowers are tubular funnel-shaped or bell-like, usually with five outward spreading lobes or points, and are often fragrant.

Species of importance to bees:

Teasel, *Dipsacus fullonum* 99

Ericaceae

Commonly known as the Heath or Heather family, the Ericaceae are usually found in acidic and infertile growing conditions. The family is large with around 4250 known species spread across 124 genera, making it the 14th most species-rich family of flowering plants. The many well-known and economically important members of the Ericaceae include the cranberry, blueberry, huckleberry, rhododendron (including azaleas), and various common heaths and heathers.

Species of importance to bees:

Ling Heather, *Calluna vulgaris* 112

Bell Heather, *Erica cinerea* 94

Cross-Leaved Heath, *Erica tetralix* 86

Winter Heath, *Erica carnea* 119

Darley Dale Heath, *Erica darleyensis* 120

Rhododendron, *Rhododendron ponticum* 51

Calico Bush, *Kalmia latifolia* 56

Blueberry, *Vaccinium corymbosum* 44

Strawberry Tree, *Arbutus unedo* 114

Pieris, *Pieris spp.* *130*

Bilberry, *Vaccinium myrtillus* *111*

Cranberry, *Vaccinium oxycoccos* *65*

Fabaceae

Worldwide, there is only one genus consisting of 150 species. The Fabacae family is a large and economically important family of flowering plants. It includes trees, shrubs and perennial or annual herbaceous plants, which are easily recognized by their fruit and their compound, stipulated leaves. The family is widely distributed and is the third-largest land plant family in terms of number of species with about 751 genera and some 19,000 known species.

Species of importance to bees:

Broad Bean, *Vicia faba* *45*

White Clover, *Trifolium repens* *69*

Red Clover, *Trifolium pratense* *68*

Sainfoin, *Onobrychis viciifolia* *75*

Bird's-Foot Trefoil, *Lotus corniculatus* *76*

Gorse, *Ulex europaeus* *123*

Broom, *Cytisus scoparius* *33*

Grossulariaceae

Worldwide, there is only 1 genus consisting of 150 species. They have regular, bisexual flowers, usually about 8 mm in diameter. The blossoms are yellow, white, pale green or sometimes red. The flowers have five united sepals and five separated petals. It matures as a berry with several to numerous seeds. They have very distinctive leaves.

Species of importance to bees:

Blackcurrant, *Ribes nigra* *43*

Redcurrant, *Ribes ruba* *36*

Gooseberry, *Ribes uva-crispa* *23*

Iridaceae

Iridaceae is a family of plants in order Asparagales, taking its name from the irises, meaning rainbow, referring to its many colours. There are 66 accepted genera with a total of circa 2244 species worldwide. It includes a number of other well-known cultivated plants.

Species of importance to bees:

Crocus, *Crocus spp.* 125

Lamiaceae

The Lamiaceae are a family of flowering plants commonly known as the mint or deadnettle family. Many of the plants are aromatic in all parts and include widely used culinary herbs. Some species are shrubs, trees (such as teak) or, rarely, vines.

Many members of the family are widely cultivated, not only for their aromatic qualities, but also their ease of cultivation, since they are rapidly propagated by stem cultivation.

Species of importance to bees:

Thyme, *Thymus spp.* 59
Mint, *Mentha spp.* 82
Lavender, *Lavandula angustifolia* 71
Rosemary, *Rosmarinus officinalis* 37
Phlomis, *phlomis* 62
Self-Heal, *Prunella grandiflora* 67
Hyssop, *Hyssopus officinalis* 102
Marjoram, *Origanum majorana* 73
Catmint, *Nepeta spp.* 78
Sage, *Salvia officinalis* 55
Lemon Balm, *Melissa officinalis* 72

Lythraceae

Lythraceae is a family of flowering plants, including 32 genera with about 620 species of herbs, shrubs and trees. Lythraceae has a worldwide distribution, with most species in the tropics, but ranging into temperate climate regions as well. The family is named after the genus, Lythrum, the loosestrifes and henna. Botanically, the leaves are usually in pairs (opposite) and the flower petals emerge from the rim of calyx tube. The petals often appear crumpled.

Species of importance to bees:

Purple Loosestrife, *Lythrum salicaria* 83

Malvaceae

Malvaceae, or the mallows, is a family of flowering plants estimated to contain 244 genera with 4225 known species. Well-known members of economic importance include okra, cotton, cacao and durian. There are also some genera containing familiar ornamentals, such as Alcea (Hollyhock), Malva (Mallow) and Lavatera (Tree Mallow).

Species of importance to bees:

Lime, *Tilia spp.*	*66*
Common Mallow, *Malva sylvestris*	*79*
Musk Mallow, *Malva moscata*	*80*
Tree Mallow, *Lavatera*	*81*
Hollyhock, *Alcea rosea*	*70*
Checkerblooms, *Sidalcea spp.*	*100*

Oleaceae

The Oleaceae are a family of flowering plants in the order Lamiales. It presently comprises 26 genera, one of which is recently extinct. The number of species in the Oleaceae is variously estimated in a wide range around 700. The Oleaceae consist of shrubs, trees, and a few lianas. The flowers are often numerous and highly odoriferous. The family has a distribustion ranging from the subarctic to the southernmost parts of Africa, Australia, and South America. Notable members of the Oleaceae include olive, ash, jasmine and several popular ornamental plants including privet, forsythia, and lilac.

Species of importance to bees:

Privet, *Ligustrum vulgare*	*74*

Onagraceae

The onagraceae are a family of flowering plants known as the Willowherb or Evening Primrose family. They include about 650 species of herbs, shrubs, and trees in 17 genera. The family is widespread, occurring on every continent from boreal and tropical regions. Some, particularly the willowherbs (epilobium), are common weeds in gardens and rapidly colonise disturbed habitats in the wild. The family is characterised by flowers with usually four sepals and petals; in some genera, such as fuschia, the sepals are as brightly coloured as the petals.

Species of importance to bees:

Rosebay Willowherb, *Chamerion angustifolium* *98*

Papaveraceae

The Papaveraceae are an economically important family of about 42 genera and approximately 775 known species of flowering plants in the order Ranunculales, informally known as the poppy family. The family is cosmopolitan, occurring in temperate and subtropical climates (mostly in the northern hemisphere), but almost unknown in the tropics. Most are herbaceous plants, but a few are shrubs and small trees.

Species of importance to bees:

Poppy, *Papaver spp.* *58*

Plumbaginaceae

Plumbaginaceae is a family of flowering plants, with a cosmopolitan distribution, The family is sometimes referred to as the leadwort family or the plumbago family. Most species in this family are perennial herbaceous plants, but a few grow as shrubs. The plants have perfect flowers and are pollinated by insects. They are found in many different climates; from arctic to tropical conditions, but are particularly associated with salt-rich marshes and sea coasts.

Species of importance to bees:

Sea Lavender, *Limonium vulgare* *108*

Ranunculaceae

Ranunculaceae is a family of over 2,000 known species of flowering plants in 43 genera, distributed worldwide. The largest are Ranunculus (600 species), Delphinium (365), Thalictrum (330), Clematis (325), and Aconitum (300). Ranunculaceae are mostly herbaceous annuals or perennials, but some are woody climbers or shrubs.

Species of importance to bees:

Clematis, *Clematis vitalba* *97*
Winter Aconite, *Eranthis hyemalis* *127*
Lenten Rose, *Helleborus orientalis* *129*

Rosaceae

A medium-sized family of flowering plants, including 4,828 known species in 91 genera. The name is derived from the type genus Rosa. Among the most species-rich genera are Alchemilla (270), Sorbus (260), Crataegus (260), Cotoneaster (260), Rubus (250), and Prunus with about 200 species. However, all of these numbers should be seen as estimates – much taxonomic work remains. The family Rosaceae includes herbs, shrubs, and trees. Most species are deciduous, but some are evergreen. They have a worldwide range but are most diverse in the Northern Hemisphere.

Species of importance to bees:

> Hawthorn, *Crataegus spp.* 27
> Cherry, *Prunus spp.* 22
> Apple, *Malus pumila* 30
> Blackberry, *Rubus fruticosus* 60
> Cherry Laurel, *Prunus laurocerasus* 34
> Cotoneaster, *Cotoneaster spp.* 46
> Raspberry, *Rubus idaeus* 84
> Pear, *Pyrus spp.* 21
> Rowan, *Sorbus spp.* 52
> Strawberry, *Fragaria spp.* 39
> Blackthorn, *Prunus spinosa* 128

Rutaceae

The Rutaceae are a family, commonly known as the rue or citrus family, of flowering plants, usually placed in the order Sapindales. Species of the family generally have flowers that divide into four or five parts, usually with strong scents. They range in form and size from herbs to shrubs and large tress. The most economically important genus in the family is Citrus, which includes the orange, lemon, grapefruit, and lime. Boronia is a large Australian genus, some members of which have highly fragrant flowers and are used in commercial oil production. About 160 genera are in the family Rutaceae.

Species of importance to bees:

> Chinese Bee Tree, *Tetradium daniellii* 95
> Mexican Orange Blossom, *Choisya ternata* 50

Salicaceae

The Salicaceae are the willow family. Recent genetics have greatly expanded the circumscription of the family to contain 56 genera and about 1220 species.

Species of importance to bees:

Willow, *Salix aegyptiaca* *126*

Willow, *Salix caprea* *25*

Sapindaceae

The Sapindaceae are a family of flowering plants in the order Sapindales, known as the soapberry family. It contains 138 genera and 1858 accepted species. The Sapindaceae occur in temperate to tropical regions, many in laurel forest habitat, throughout the world. Many are laticiferous, i.e., they contain latex, a milky-sap, and may contain milky toxic saponin with soap-like qualities in wither the foliage, seeds and/or roots.

Species of importance to bees:

Horse Chestnut, *Aesculus hippocastanum* *48*

Maple, *Acer spp.* *49*

Sycamore, *Acer pseudoplatanus* *38*

Verbenaceae

The Verbenaceae are a family, commonly known as the verbena family or vervain family, of mainly tropical flowering plants. It contains trees, shrubs, and herbs notable for heads, spikes, or clusters of small flowers, many of which have an aromatic smell. The Verbenaceae family includes around 35 genera and 1200 species. Economically important Verbenaceae include lemon verbena (*Aloysia triphylla*), grown for aroma or flavouring, and verbenas or vervains (verbena), some of which are used in herbalism and others grown in gardens.

Species of importance to bees:

Verbena, *Verbena bonariensis* *109*

Flowering Periods

Plant Families

Species	M	A	M	J	J	A	S	O	N	D	J	F
Pear	•											
Cherry	•	•										
Gooseberry	•	•										
Lungwort	•	•										
Pussy Willow	•	•										
Aubretia	•	•	•									
Hawthorn	•	•	•	•								
Dandelion	•	•	•	•	•	•	•	•				
Apple		•										
Barberry		•	•									
Bluebell		•	•									
Broom		•	•									
Cherry Laurel		•	•									
Oilseed Rape		•	•									
Redcurrant		•	•									
Rosemary		•	•	•								
Sycamore		•	•	•								
Strawberry		•	•	•	•							
Forget-Me-Not		•	•	•	•	•						
Asparagus			•	•								
Blackcurrant			•	•								
Blueberry			•	•								
Broad Bean			•	•								
Cotoneaster			•	•								
Holly			•	•								
Horse Chestnut			•	•								
Maple			•	•								
Mexican Orange			•	•								
Rhododendron			•	•								
Rowan			•	•								
Comfrey			•	•	•							
Lovage			•	•	•							
Sage			•	•	•							
Calico Bush			•	•	•	•						

Species	M	A	M	J	J	A	S	O	N	D	J	F
Charlock			•	•	•	•						
Poppy			•	•	•	•						
Thyme			•	•	•	•						
Blackberry			•	•	•	•	•					
Phacelia			•	•	•	•	•					
Phlomis			•	•	•	•	•					
Vipers Bugloss			•	•	•	•	•					
Cranberry				•	•							
Lime				•	•							
Self-Heal				•	•							
Clover, Red				•	•	•						
Clover, White				•	•	•						
Hollyhock				•	•	•						
Lavender				•	•	•						
Lemon Balm				•	•	•						
Marjoram				•	•	•						
Privet				•	•	•						
Sainfoin				•	•	•						
Bird's-Foot Trefoil				•	•	•	•					
Borage				•	•	•	•					
Catmint				•	•	•	•					
Mallow, Common				•	•	•	•					
Mallow, Musk				•	•	•	•					
Mallow, Tree				•	•	•	•					
Mint				•	•	•	•					
Purple Loosestrife				•	•	•	•					
Raspberry				•	•	•	•					
Wild Carrot				•	•	•	•					
Cross-Leaved Heath				•	•	•	•	•				
Himalayan Balsam				•	•	•	•	•				
Sweet Alison				•	•	•	•	•				
Dahlia				•	•	•	•	•	•	•		
Garden Angelica				•								
Rudbeckia				•								

Plant Families

19

Plant Families

Species	M	A	M	J	J	A	S	O	N	D	J	F
Yarrow					•							
Bell Heather					•	•						
Chinese Bee Tree					•	•						
Chives					•	•						
Clematis					•	•						
Rosebay Willowherb					•	•						
Teasel					•	•						
Checkerbloom					•	•	•					
Golden Rod					•	•	•					
Hyssop					•	•	•					
Knapweed					•	•	•					
Sunflower					•	•	•					
Tansy					•	•	•					
Ragwort					•	•	•	•				
Sea Aster					•	•	•	•				
Sea Lavender					•	•	•	•				
Verbena					•	•	•	•	•			
Bilberry						•	•					
Ling Heather						•	•	•				
Strawberry Tree							•	•	•			
Ivy							•	•	•			
Mahonia									•	•	•	•
Winter Heath	•									•	•	•
Darley Dale Heath	•	•	•							•	•	•
Snowdrop	•										•	•
Gorse	•	•	•	•							•	•
Crocus	•											•
Musk Willow	•											•
Winter Aconite	•											•
Blackthorn	•	•										•
Lenten Rose	•	•										•
Pieris	•	•										•

Plants by Main Flowering Periods

SPRING

Pear

Family: Rosaceae

Common Name: Pear

Latin Name: *Pyrus spp.*

Description: Fruit-bearing tree up to 17m tall

Pollen: Light green 35μm diameter

Nectar: No honey crop

Leaves: Simple, 2-12cm long, glossy green on some species, densely silvery-hairy in some others; leaf shape varies from broad oval to narrow lanceolate.

Flowers: White, occasionally tinted yellow or pink, 2-4cm diameter, and with five petals.

Flowering times: March

Cherry

Family: Rosaceae

Common Name: Cherry

Latin Name: *Prunus spp.*

Description: A genus of trees and shrubs

Pollen: Orange-brown 35µm diameter

Nectar: Minor honey crop. Pale yellow colour. Rapid granulation

Other Names: Stone Fruit

Leaves: Simple, alternate, usually lanceolate, unlobed, and often with nectaries on the leaf stalk

Flowers: are usually white to pink, sometime red, with five petals and five sepals. There are numerous stamens. Flowers are borne singly, or in umbels of two to six or sometimes more on racemes.

Flowering times: March – April

GOOSEBERRY

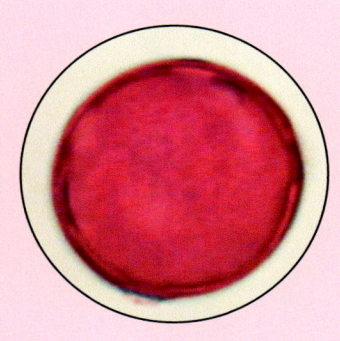

Family: Grossulariaceae (currants)

Common Name: Gooseberry

Latin Name: *Ribes uva-crispa*

Description: Bush producing an edible fruit

Pollen: Yellow 30 μm diameter

Nectar: No honey crop

Leaves: Groups of rounded, deeply crenated 3 or 5 shallow-lobed and dark green leaves. It can be prone to a leaf-spot disease that can eventually weaken the plant and shivel the fruit.

Flowers: Gooseberry differs from currants in that their flowers grow one to three together on short stems, not in racemes. The flowers are bell-shaped with five petals and are usually red to green in colour.

Flowering times: March – April

Lungwort

Family: Boraginaceae

Common Name: Lungwort

Latin Name: *Pulmonaria officinalis*

Description: A rhizomatous evergreen perennial

Pollen: White, 30μm diameter

Nectar: No honey crop

Other Names: Mary's Tears, Our Lady's Milk Drops, Common Lungwort

Leaves: Basal leaves are green, cordate, elongated and pointed. They always have rounded and often sharply defined white or pale-green patches

Flowers: The five-petal flowers are red or pink at first, later turn to blue-purple.

Flowering Times: March – April

Pussy Willow

Family: Salicaceae

Common Name: Pussy Willow

Latin Name: *Salix caprea*

Description: Deciduous shrub and tree

Pollen: Yellow 20μm diameter

Nectar: Not a significant crop. Light golden yellow. Mild flavour, fine granulation

Other Names: Goat Willow, Great Sallow

Leaves: Oval rather than long and thin. They are hairless above, with a felty coating of fine grey hairs underneath and a pointed tip which bends to one side

Flowers: Male catkins are grey stout and oval, becoming yellow when ripe with pollen. Female catkins are longer and green. Pussy Willows are dioecious

Flowering Times: March – April

AUBRETIA

Family: Brassicaceae

Common Name: Aubretia

Latin Name: *Aubretia deltoidea*

Description: A small herbaceous perennial

Pollen: Light green with a 25μm diameter

Nectar: No honey crop

Other Names: Lilacbush, Purple Rock Cress, Rainbow Rock Cress

Leaves: Green and spoon to oval-shaped, some of which are lobed.

Flowers: The showy inflorescence bears small flowers with four lavender to deep pink petals.

Flowering Times: March – May

HAWTHORN

Family: Rosaceae

Common Name: Hawthorn

Latin Name: *Crataegus spp.*

Description: A large genus of shrubs and trees

Pollen: Green 40μm diameter

Nectar: No honey crop, except rarely - the conditions have to be perfect

Other Names: Thornapple, May-tree, Whitethorn or Hawberry

Leaves: highly variable, but generally alternate, simple, 50-100mm long, serrate and lobed (may be unlobed), long thorns, dark green above and paler below

Flowers: Perfect, usually small white flowers, with five petals produced in clusters near the end of the twig

Flowering times: March – June

Mar - June

Dandelion

Family: Asteraceae

Common Name: Dandelion

Latin Name: *Taraxacum officinale*

Description: A herbaceous perennial plant

Pollen: Orange, 30μm diameter

Nectar: A minor honey crop. Intense golden yellow colour with rapid, coarse, hard granulation. The flavour is strong and sharp with an aroma that is initially repellent, like the flower.

Leaves: The leaves are 4-45cm long and 1-10cm wide. They are oblanceolate, oblong or ovate in shape with jagged edges.

Flowers: The yellow flower heads lack receptacle bracts and all the flowers, which are called florets, are ligulate and bisexual

Flowering Times: March – October

APPLE

Family: Rosaceae

Common Name: Apple

Latin Name: *Malus pumila*

Description: A deciduous tree in the rose family

Pollen: Yellow-brown 40µm diameter

Nectar: Minor honey crop. Amber colour with an aroma of apples. Granulation irregular

Leaves: Alternately arranged dark green-coloured simple ovals with serrated margins and slightly downy undersides.

Flowers: the 3-4cm flowers are white with a pink tinge that gradually fades, five petaled, with an inflorescence consisting of a cyme with 4-6 flowers.

Flowering times: from April depending on variety

BARBERRY

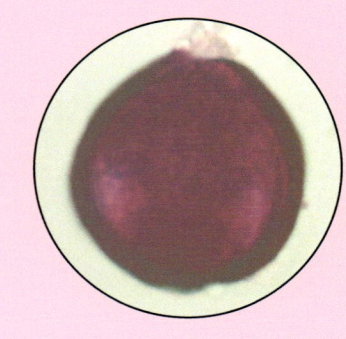

Family: Berberidaceae

Common Name: Barberry

Latin Name: *Berberis spp.*

Description: A large genus of shrubs

Pollen: Yellow-Orange, 35µm diameter

Nectar: No honey crop

Leaves: The genus Berberis has dimorphic shoots: The leaves on long shoots are non-photosynthetic, developed into spines 3–30 mm long. A short shoot develops between these thorn-leaves with several photosynthetic leaves; 1–10cm long, simple, and either entire, or with spiny margins.

Flowers: The flowers are produced singly or in racemes of up to 20 on a single flower-head. They are yellow or orange, 3-6mm long

Flowering Times: April – May

Bluebell

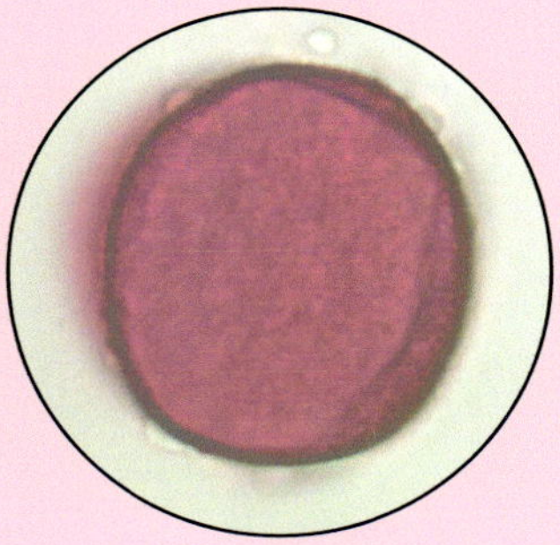

Family: Asparagaceae

Common Name: Bluebell

Latin Name: *Hyacinthoides non-scripta*

Description: Bulbous perennial woodland plant

Pollen: Blue, 40x50µm

Nectar: No honey crop

Other Names: Common Bluebell, Harebell

Leaves: The plant produces 3-6 linear leaves, all growing from the base of the plant, and each 7-16mm wide.

Flowers: An inflorescence of 5-12 (exceptionally 3-32) flowers is borne on a stem up to 500mm tall, which droops towards the tip the flowers are arranged in a one-sided, nodding raceme. Each flower is 14-20mm long.

Flowering Times: April - May

Apr - May

Broom

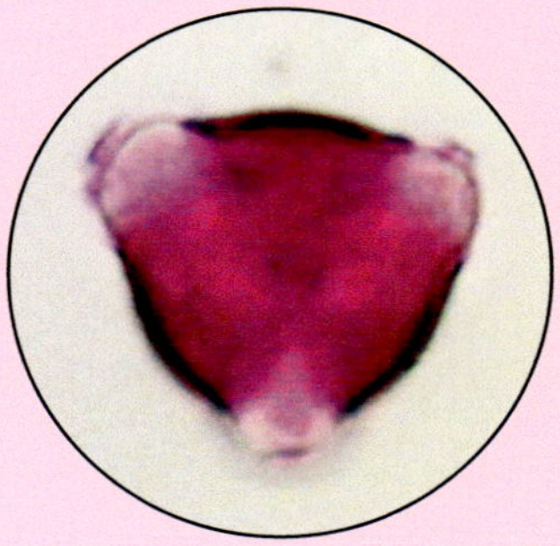

Family: Fabaceae

Common Name: Broom

Latin Name: *Cytisus scoparius*

Description: A perennial leguminous shrub

Pollen: Orange 25 μm diameter

Nectar: No honey crop

Other Names: Common broom or Scotch broom

Leaves: the shrubs have green shoots with small deciduous triloliate leaves 5-15mm long

Flowers: in spring and summer is covered in profuse golden yellow flowers 20-30mm from top to bottom and 15-20mm wide

Flowering times: April – May

Cherry Laurel

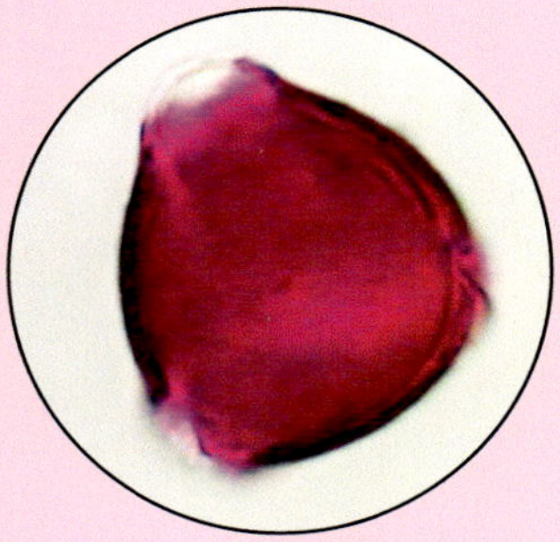

Family: Rosaceae

Common Name: Cherry Laurel

Latin Name: *Prunus laurocerasus*

Description: An evergreen species of cherry

Pollen: Dull green, 40µm diameter

Nectar: No honey crop

Other Names: Common Laurel, English Laurel

Leaves: Dark green, leathery, shiny, 10-25cm long and 4-10cm broad, with a finely serrated margin. The leaves can have the scent of almonds when crushed

Flowers: Erect 7-5cm racemes of 30-40 flowers, each flower 1cm across, with five creamy-white petals and numerous yellowish stamens with a sweet smell.

Flowering Times: April - May

Apr - May

OILSEED RAPE

Family: Brassicaceae

Common Name: Oilseed Rape

Latin Name: *Brassica napus*

Description: A yellow-flowering field crop

Pollen: Light yellow-green 30μm diameter

Nectar: Major honey crop. Pale with a very rapid granulation. Mild and sweet.

Other Names: Rapeseed, Rape, OSR

Leaves: Lanceolate, bluish-green leaves grown alternately, the basal leaves are stalked and shallow-lobed, but the upper leaves are stalkless and entire.

Flowers: The yellow flowers have four petals alternating with four sepals and two lateral stamens with short filaments, and four median stamens with longer filaments.

Flowering Times: April – May

REDCURRANT

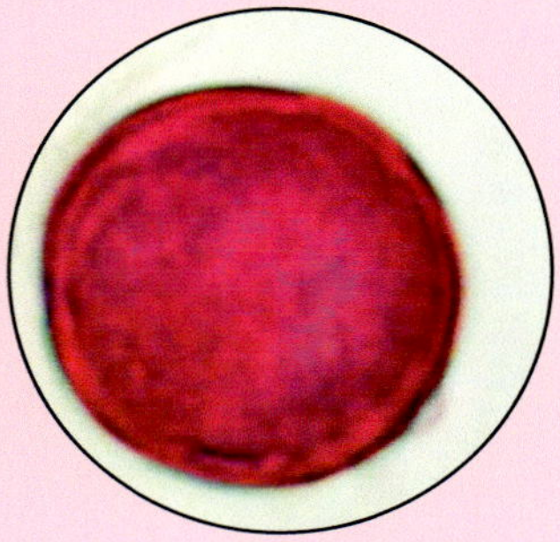

Family: Grossulariaceae (currants)

Common Name: Redcurrants

Latin Name: *Ribes rubra*

Description: A shrub normally growing to 1-1.5m

Pollen: Yellow 30 μm diameter

Nectar: No honey crop

Other Names: Non

Leaves: Five-lobed leaves arranged spirally around the stem

Flowers: The flowers are inconspicuous yellow-green, in pendulous 4-8cm drooping racemes

Flowering times: April – May

Apr - May

Rosemary

Family: Lamiaceae

Common Name: Rosemary

Latin Name: *Rosmarinus officinalis*

Description: A perennial, fragrant, evergreen herb

Pollen: White, 40μm diameter

Nectar: Not a major honey crop in the UK. Light and clear with a distinctive aroma and flavour. It usually has rapid, fine granulation.

Other Names: Anthos

Leaves: Growing between 2-4cm long and 2-5mm broad, the leaves are evergreen with dense woolly hair, coloured green above and pale-green to white below.

Flowers: Its flowers are white, pink, purple or deep blue. Rosemary also has a tendency to flower outside of its usual season.

Flowering Times: April – June

Sycamore

Family: Sapindaceae

Common Name: Sycamore

Latin Name: *Acer pseudoplatanus*

Description: A large deciduous, broad-leaved tree.

Pollen: Yellow-green 40µm diameter

Nectar: Medium honey crop. Light amber colour with fine granulation. Rank flavour that improves with age.

Leaves: Grow on long leafstalks and are large and palmate, with five large radiating lobes.

Flowers: Greenish-yellow and they hang in dangling flowerheads called panicles. They produce copious amounts of pollen and nectar.

Flowering Times: April - June

Strawberry

Family: Rosaceae

Common Name: Strawberry

Latin Name: *Fragaria spp.*

Description: Herbaceous perennial plant 10-15cm tall

Pollen: Yellow-dark green 20μm diameter

Nectar: No honey crop

Leaves: Consisting of several basal leaves and one or more inflorescences. The basal leaves are trifoliate

Flowers: Each flower has five white petals, five green sepals and five green sepal-like bracts. They can be pistillate, staminate, or perfect.

Flowering times: April – July

FORGET-ME-NOT

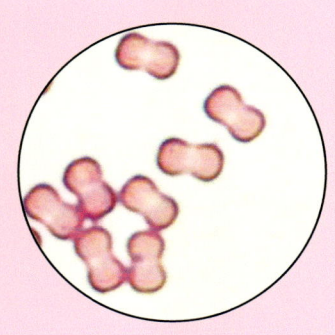

Family: Boraginaceae

Common Name: Forget-Me-Not

Latin Name: *Myotosis spp.*

Description: Annual or perennial flowering plants

Pollen: Yellow-orange and oval-shaped, 2 x 5μm in size

Nectar: No honey crop

Other Names: Scorpion Grasses

Leaves: Alternate, simple and entire.

Flowers: These are typically 1cm diameter (or less), flat and blue, pink, white or yellow with yellow centres, growing on scorpioid cymes.

Flowering Times: April – September

Apr - Sept

Asparagus

Family: Asparagaceae

Common Name: Asparagus

Latin Name: *Asparagus officinalis*

Description: A spring vegetable, perennial plant

Pollen: Orange, 25μm diameter

Nectar: No honey crop

Other Names: Garden Asparagus, Sparrow Grass

Leaves: They are in fact needle-like cladodes (modified stems) in the axils of scale leaves; they are 6-32mm long and 1mm broad.

Flowers: Flowers are bell-shaped, greenish-white to pale yellow, that grows to 4.5-6.5mm long. Usually dioecious.

Flowering Times: May - June

BLACKCURRANT

Family: Grossulariaceae (currants)

Common Name: Blackcurrant

Latin Name: *Ribes nigrum*

Description: A medium-sized shrub

Pollen: 30 µm diameter

Nectar: No honey crop

Leaves: Alternate, with fairly long stalks, strong-scented. Blade with palmate venation, 3-5 lobed, with cordate base, toothed margins.

Flowers: inconspicuous, regular, approximately 8mm across, reddish or brownish green. Calyx five-lobed, wheel-shaped, hairy with yellow glands.

Flowering times: May – June

Blueberry

Family: Ericaceae

Common Name: Blueberry

Latin Name: *Vaccinium corymbosum*

Description: A medium-sized deciduous shrub with small blue fruit

Pollen: Pink, 30μm diameter

Nectar: No honey crop.

Other Names: Blue Huckleberry, Tall Huckleberry, Swamp Huckleberry, High Huckleberry

Leaves: The dark, glossy green leaves are elliptical and up to 5cm long

Flowers: White to very light pink, the flowers are bell or urn-shaped, growing to 8.4mm long

Flowering Times: May - June

BROAD BEAN

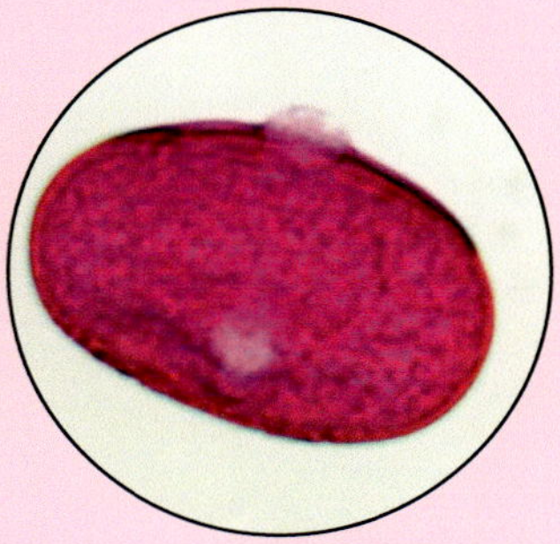

Family: Fabaceae

Common Name: Broad Bean

Latin Name: *Vicia faba*

Description: A stiffly erect plant 0.5-1.8m tall

Pollen: Grey-green 40x20 μm oblong

Nectar: Major honey crop: mild flavour, light colour, rapid coarse granulation

Other Names: Field Bean

Leaves: The leaves are 10-25cm long, pinnate with 2-7 leaflets, and of a distinct glaucous grey-green colour

Flowers: The flowers are 1-2.5cm long, with five petals – white standard petals and wing petal white with black spot

Flowering times: May – June

May -June

COTONEASTER

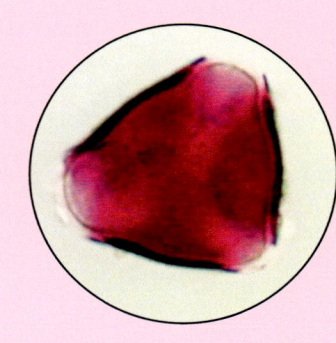

Family: Rosaceae

Common Name: Cotoneaster

Latin Name: *Cotoneaster spp.*

Description: A genus of flowering plants

Pollen: Yellow-green, 30μm diameter

Nectar: No honey crop.

Leaves: Arranged alternately, 0.5-15cm long, ovate to lanceolate in shape and entire. Plants may be evergreen or deciduous.

Flowers: Solitary or in corymbs of up to 100 together. The flower is either fully open or has five petals half open, 5-10mm in diameter. They may be any shade from white, through to pink to red.

Flowering Times: May - June

Holly

Family: Aquifoliaceae

Common Name: Holly

Latin Name: *Ilex aquifolium*

Description: An evergreen tree or shrub

Pollen: Yellow-green, 35µm diameter

Nectar: No honey crop.

Other Names: Common Holly, English Holly, European Holly, Christmas Holly

Leaves: The leaves are leathery and shiny, growing 5-12cm long and 2-6cm broad. They are evergreen, a dark green on the upper surface and lighter on the underside.

Flowers: Holly is dioecious. In male specimens, the flowers are yellowish and appear in axillary groups. In the female, the small flowers are isolated or in groups of three and coloured white or slightly pink.

Flowering Times: May – June

Horse Chestnut

Family: Sapindaceae

Common Name: Horse Chestnut

Latin Name: *Aesculus hipposcastanum*

Description: A large tree, growing up to 40m tall

Pollen: Red-brown 20µm diameter

Nectar: Not a significant honey crop. Dark with rapid granulation

Other Names: Conker Tree

Leaves: Opposite and palmately compound, with 5 to 7 leaflets, each 13 – 30cm long, making the whole leaf up to 60cm across.

Flowers: Usually white with a yellow to pink blotch at the base of the petals. In erect panicles 10 – 30cm tall with about 20 to 50 flowers on each panicle.

Flowering Times: May – June

Maple

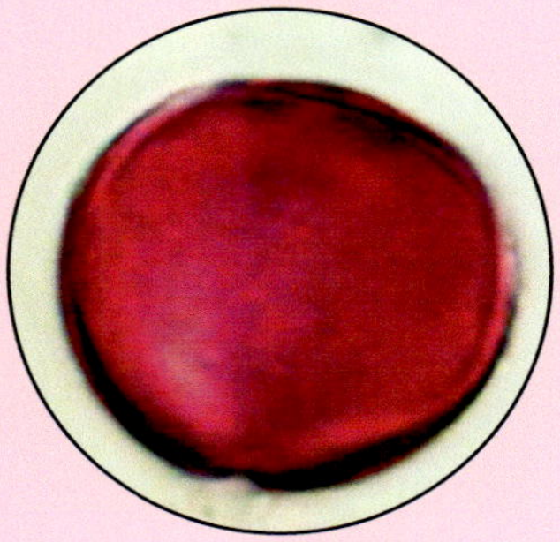

Family: Sapindaceae

Common Name: Maple

Latin Name: *Acer spp.*

Description: A genus of trees or shrubs

Pollen: Light to dark green 40μm diameter

Nectar: No honey crop.

Leaves: Palmate veined and lobed, with 3 – 9 (in rare cases to 13) veins each leading to a lobe.

Flowers: Regular, pentamerous, and borne in racemes, corymbs or umbels. Four or five petals about 1 – 6mm long.

Flowering Times: May – June

May - June

Mexican Orange Blossom

Family: Rutaceae

Common Name: Mexican Orange Blossom

Latin Name: *Choisya ternata*

Description: Evergreen shrub, growing up to 3m

Pollen: Yellow, 30μm diameter

Nectar: No honey crop

Other Names: Mexican Orange, Choisya

Leaves: The palmate leaves are made up of three dark-green leaflets that release a citrus aroma when rubbed.

Flowers: The white scented flowers appear in spring, occasionally with a limited repeat flowering in autumn

Flowering Times: May – June

RHODODENDRON

Family: Ericaceae

Common Name: Rhododendron

Latin Name: *Rhododendron ponticum*

Description: An invasive dense, suckering shrub or small tree, a non-native species.

Pollen: White, 30μm diameter

Nectar: No honey crop – toxic to bees and possibly humans. The nectar's grayanotoxins cause palpitations, paralysis and death within hours for honeybees.

Leaves: The leaves are evergreen, growing 6 – 18cm long and 2 – 5cm wide in an elliptical shape. They are dark green and glossy on top with waxy or leathery feel.

Flowers: The flowers are a pale purple, often with greenish-yellow spots or streaks. They grow to between 3.5 – 5cm in diameter and have five, slightly crinkled petals.

Flowering Times: May - June

ROWAN

Family: Rosaceae

Common Name: Rowan

Latin Name: *Sorbus spp.*

Description: Small deciduous trees 10-20m tall

Pollen: Light green 30µm diameter

Nectar: No honey crop

Other Names: Whitebeam, Service Tree, Mountain-Ash

Leaves: Pinnate comprising 5-8 pairs of leaflets, plus one 'terminal' leaflet. Each leaflet is long, oval and toothed.

Flowers: Borne in dense clusters, each one bearing five creamy white petals.

Flowering times: May – June

COMFREY

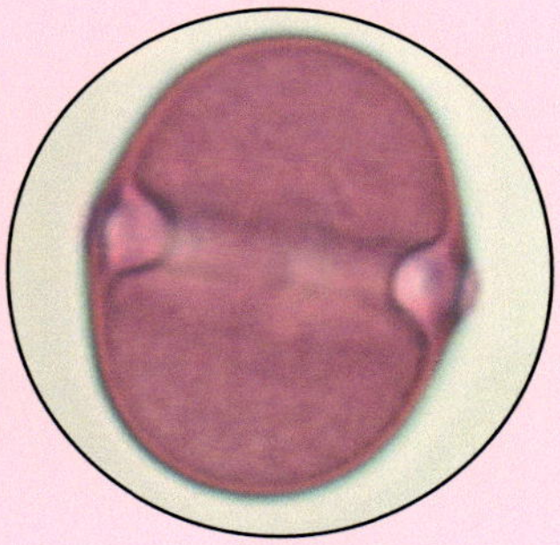

Family: Boraginaceae

Common Name: Comfrey

Latin Name: *Symphytum spp.*

Description: A robust perennial forming a clump of erect stems

Pollen: White-yellow, 20x30μm in size

Nectar: No honey crop

Other Names: Comphrey

Leaves: Large, hairy, broad leaves with pointed ends

Flowers: Small, bell-shaped flowers of various colours, typically cream or purple, which may be striped.

Flowering Times: May – July

LOVAGE

Family: Apiaceae

Common Name: Lovage

Latin Name: *Levisticum officinale*

Description: Erect, herbaceous, perennial plants

Pollen: Yellow, 30μm diameter

Nectar: No honey crop

Leaves: A basal rosette of leaves and stems with further leaves. The larger basal leaves are up to 70cm long.

Flowers: The flowers are yellow to greenish-yellow, 2-3mm in diameter, produced in globose umbels up to 10-15cm diameter at the top of the stems.

Flowering Times: May – July

SAGE

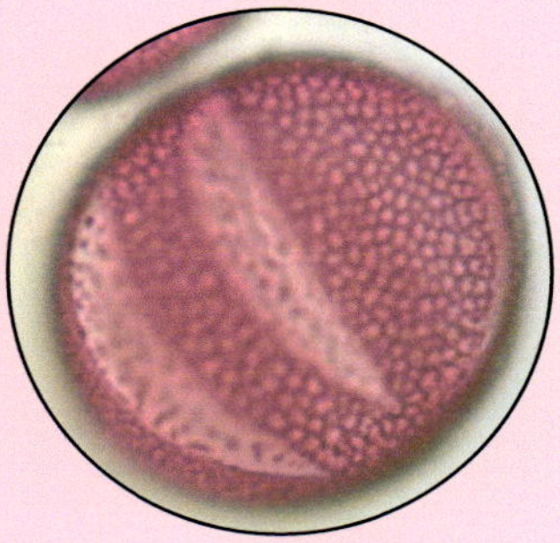

Family: Lamiaceae

Common Name: Sage

Latin Name: *Salvia officinalis*

Description: Perennial, woody evergreen shrub

Pollen: Yellow, 45μm diameter

Nectar: Not a significant crop. Pale yellow to amber in colour with an aromatic flavour and slow granulation.

Other Names: Garden Sage, Common Sage, Culinary Sage

Leaves: The leaves are oblong, ranging from in size up to 6.5cm long by 2.5cm wide. They are grey-green, rugose on the upper side and nearly white underneath

Flowers: Most commonly, they have lavender-coloured flowers but they can also be white, pink or purple.

Flowering Times: May – July

Calico Bush

Family: Ericaceae

Common Name: Calico Bush

Latin Name: *Kalmia latifolia*

Description: A broad-leaved, dense, evergreen shrub

Pollen: Yellow, 40μm diameter

Nectar: No honey crop – poisonous to bees and humans

Other Names: Mountain Laurel or Spoonwood

Leaves: The leaves are 3 – 12cm long and 1 – 4cm wide

Flowers: The flowers have a round, umbrella-liked shape occurring in clusters. Their colour usually ranges from white to light pink.

Flowering Times: May - August

CHARLOCK

Family: Brassicaceae

Common Name: Charlock

Latin Name: *Sinapis arvensis*

Description: An annual or winter annual plant

Pollen: Light yellow-green 30μm diameter

Nectar: Not a significant crop. Light to amber colour. Rapid coarse granulation. Mild in flavour with a slight sharpness similar to mustard.

Other Names: Charlock Mustard, Field Mustard, Wild Mustard

Leaves: Petiolate (stalked with a length of 1 – 4cm.

Flowers: The inflorescence is a raceme made up of yellow flowers which each have four petals.

Flowering Times: May – August

Poppy

Family: Papaveraceae (Poppy Family)

Common Name: Poppy

Latin Name: *Papaver spp.*

Description: Annuals with 1-2 pinnate leaves (having leaflets arranged on either side of the stem, typically in pairs opposite each other)

Pollen: Dark blue/Black 25 µm diameter

Nectar: No honey crop

Leaves: Poppies have lobed or dissected leaves

Flowers: Flowers of species have 4 or 6 petals, many stamens forming a conspicuous whorl in the centre of the flower. The petals are showy, may be of almost any colour and some have markings.

Flowering times: May – August

Thyme

Family: Lamiceae

Common Name: Thyme

Latin Name: *Thymus spp.*

Description: 350 species of aromatic perennial shrubs.

Pollen: Yellow-orange, 35µm diameter

Nectar: A minor honey source. Amber in colour with a strong, fragrant aroma and a strong, minty flavour which does not fade on storage. Slow granulation

Leaves: In most species, they are evergreen, arranged in opposite paris, oval, entire and small, reaching just 4-20mm long, and usually aromatic.

Flowers: Produced in dense, terminal heads, with an uneven calyx and a three-lobed upper lip. They are usually yellow, white or purple.

Flowering Times: May – August

Blackberry

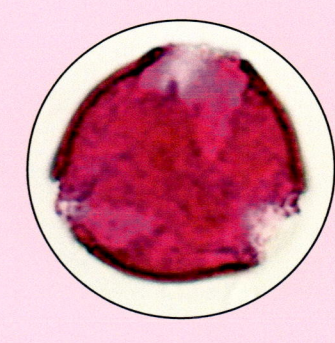

Family: Rosaceae

Common Name: Blackberry

Latin Name: *Rubus fruticosus*

Description: An edible fruit producing shrub

Pollen: Dull green 30μm diameter

Nectar: Important honey crop. Light colour with a delicate flavour like clover. Slow granulation.

Other Names: Bramble

Leaves: Large palmately compound leaves with five or seven leaflets

Flowers: produced in late spring and early summer on short racemes on the tips of the flowering laterals. Each flower is about 2-3cm in diameter with five white or pale pink petals.

Flowering times: May – September

Phacelia

Family: Boraginaceae

Common Name: Phacelia

Latin Name: *Phacelia tanacetifolia*

Description: An annual herb which grows erect

Pollen: Dark blue, 20µm diameter

Nectar: Not considered a major honey crop. Colour is white to amber with a mild flavour and rapid granulation.

Other Names: Lacy Phacelia, Blue Tansy, Purple Tansy

Leaves: The leaves are mostly divided into smaller leaflets deeply and intricately cut into toothed lobes, giving them a lacy appearance.

Flowers: A one-sided curving or coiling cyme of bell-shaped flowers in shades of blue and lavender. Each flower is almost 1cm long with protruding whiskery stamens.

Flowering Times: May – September

Phlomis

Family: Lamiaceae

Common Name: Phlomis

Latin Name: *Phlomis fruticosa*

Description: A small, spreading, evergreen shrub

Pollen: Yellow, 40µm diameter

Nectar: No honey crop

Other Names: Jerusalem Sage, Lampwick Plant, Jupiter's Distaff, Yellow Clary

Leaves: The leaves are entire, opposite and decussate. They may be rugose or reticulate veined.

Flowers: They are arranged in whorls called verticillasters, which encircle the stems. The colour of the flowers varies from yellow to pink, purple and white.

Flowering Times: May – September

VIPERS BUGLOSS

Family: Boraginaceae

Common Name: Vipers Bugloss

Latin Name: *Echium vulgare*

Description: A biennial or monocarpic perennial

Pollen: Dark blue-black, 10-15µm diameter

Nectar: Produces large amounts of nectar but not a significant honey crop. It is light-coloured, with pleasant smell and delicate taste. However, large quantities of this honey may be toxic to humans.

Other Names: Blueweed, Snake Flower, Blue Borage

Leaves: It has rough, hairy, oblanceolate leaves

Flowers: The flowers start pink and turn vivid blue with protruding red stamens. Each flower grows 15-20mm long and are produced in branched spikes.

Flowering Times: May – September

May -Sept

CRANBERRY

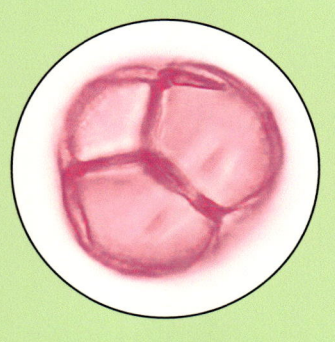

Family: Ericaceae

Common Name: Cranberry

Latin Name: *Vaccinium oxycoccos*

Description: A species of flowering plant with red, edible fruit

Pollen: White-grey, 30µm diameter

Nectar: No honey crop

Other Names: Small Cranberry, Bog Cranberry, Swamp Cranberry

Leaves: They are leathery and lance-shaped, up to 1cm long.

Flowers: Arising on nodding stalks a few centimetres tall, the corolla is white or pink and flexed backwards away from the centre of the flower.

Flowering Times: June – July

LIME

Family: Malvaceae

Common Name: Lime

Latin Name: *Tilia spp.*

Description: 30 species of trees or bushes

Pollen: Yellow, 30μm diameter

Nectar: Can be a major honey crop. Very light colour with slow granulation. Prone to honeydew which tastes like treacle. Bees have been found dead or drunk under trees.

Leaves: Dark green in colour, heart-shaped and flimsy and measure 6 – 10cm in length

Flowers: White-yellow, five-petalled and hang in clusters of 2 to 5. They have a drooping habit.

Flowering Times: June – July

Self-Heal

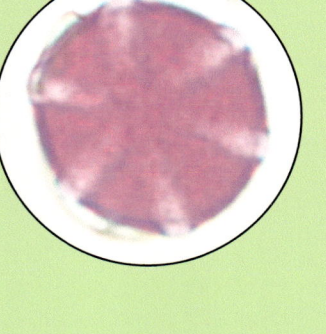

Family: Lamiaceae

Common Name: Self-Heal

Latin Name: *Prunella grandiflora*

Description: Low-growing perennial

Pollen: White, 50μm diameter

Nectar: No honey crop.

Other Names: Large Self-Heal, Large-Flowered Self-Heal

Leaves: A pretty, ground cover plant that produces a thick carpet of deep green foliage with a noticeable gap between the uppermost leaf-pair and the base of the flower-head.

Flowers: Short spikes of hooded white to blue flowers

Flowering Times: June – July

Red Clover

Family: Fabaceae

Common Name: Red Clover

Latin Name: *Trifolium pretense*

Description: A herbaceous, short-lived perennial

Pollen: Brown 30 μm diameter

Nectar: Medium honey crop, mild flavour, light amber colour, rapid granulation

Leaves: The leaves are alternate, trifoliate (with three leaflets), each leaflet 15-30mm long and 8-15mm broad.

Flowers: The flowers are light pink with a yellowish base, 10-15mm long, produced in a dense inflorescence

Flowering times: June – August

White Clover

Family: Fabaceae

Common Name: White Clover

Latin Name: *Trifolium repens*

Description: Herbaceous perennial plant

Pollen: Dull green 30 μm diameter

Nectar: Major honey crop, mild flavour with delicate after-taste, Light colour. Slow fine granulation

Leaves: The leaves are trifoliate, smooth, elliptic to egg-shaped and long-petioled and usually with light or dark markings

Flowers: Heads of whitish flowers, often with a tinge of pink or cream that may come on with the aging of the plant. The heads are generally 1.5-2cm wide

Flowering times: June - August

HOLLYHOCK

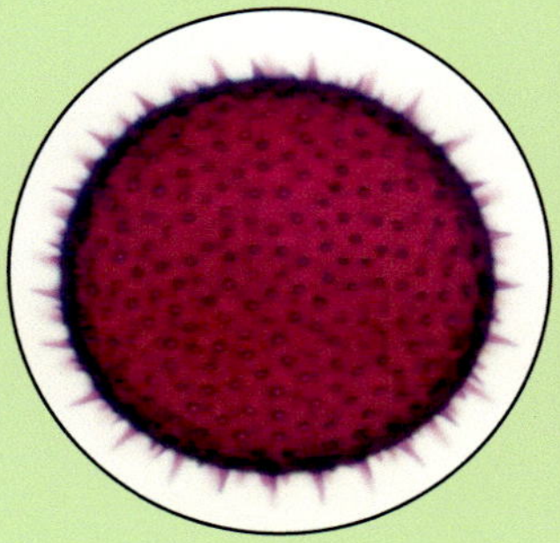

Family: Malvaceae

Common Name: Hollyhock

Latin Name: *Alcea rosea*

Description: Annual, biennial or perennial plants

Pollen: Brown 100μm diameter

Nectar: No honey crop

Leaves: The deep green leaf blades are often lobed or toothed, and are borne on long petioles (stalks).

Flowers: Solitary or arranged in fascicles or racemes. The notched petals are usually over 3cm wide and may be pink, white, purple or yellow.

Flowering Times: June – August

Lavender

Family: Lamiaceae

Common Name: Lavender

Latin Name: *Lavandula angustifolia*

Description: Aromatic shrub growing to 2m **Pollen:** Orange, 10μm diameter

Nectar: Not considered a major honey crop. Light in colour, with a fragrant aroma and fine granulation.

Other Names: True Lavender, English Lavender, Garden Lavender, Common Lavender, Narrow-Leaved Lavender.

Leaves: The leaves are evergreen, 2-6cm long and 4-6mm broad.

Flowers: The lavender purple flowers are produced on spikes 2-8cm long at the top of slender, leafless stems that are 10-30cm tall.

Flowering Times: June – August

Lemon Balm

Family: Lamiaceae

Common Name: Lemon Balm

Latin Name: *Melissa officinalis*

Description: A perennial herbaceous plant

Pollen: White-grey, 40μm diameter

Nectar: No honey crop

Other Names: Balm, Common Balm, Balm Mint

Leaves: Opposite pairs of toothed, ovate leaves growing on square, branching stems. The leaves have a mild menthol-lemon scent

Flowers: Small, white flowers that are full of nectar appear. White or light pink flowers arranged in axillary whorls.

Flowering Times: June – August

Marjoram

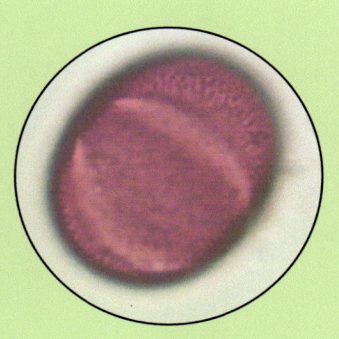

Family: Lamiaceae

Common Name: Marjoram

Latin Name: *Origanum majorana*

Description: A cold-sensitive perennial herb

Pollen: Yellow-orange, 35μm diameter

Nectar: No honey crop

Other Names: Pot Marjoram, Oregano

Leaves: Produced on square stems that are densely clad with ovate, highly aromatic, pubescent, grey-green leaves 3cm long.

Flowers: Tiny, two-lipped, tubular, white or pale pink flowers with grey-green bracts bloom in spike-like clusters.

Flowering Times: June – August

PRIVET

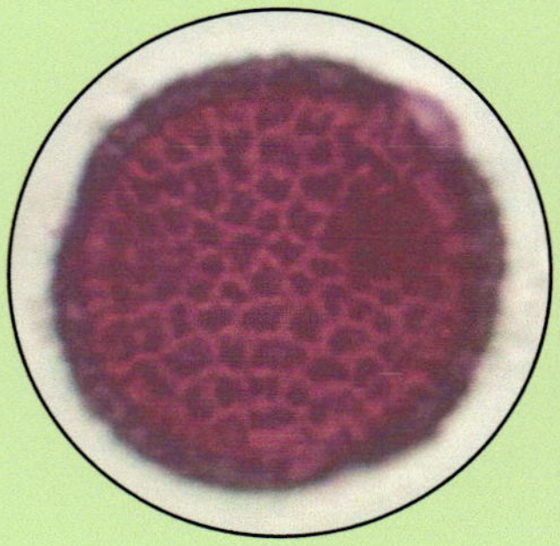

Family: Oleaceae

Common Name: Privet

Latin Name: *Ligustrum vulgare*

Description: Common hedging shrub

Pollen: Yellow, 30μm diameter

Nectar: No honey crop

Other Names: Wild Privet, Common Privet, European Privet

Leaves: The leaves are borne in opposite pairs. They are sub-shiny green and narrow oval to lanceolate. They grow 2-6cm long and 0.5-1.5cm broad.

Flowers: The flowers are produced in panicles that are 3-6cm long, each flower creamy-white, with tubular base and a four-lobed corolla 'petals' 4-6m diameter. Produce a strong, pungent fragrance that many people find unpleasant.

Flowering Times: June – August

Sainfoin

Family: Fabaceae

Common Name: Sainfoin

Latin Name: *Onobrychis viciifolia*

Description: A deep-rooted perennial legume

Pollen: Dark brown 20x40µm oblong

Nectar: Medium honey crop, sweet pronounced flavour, pale yellow colour, rapid coarse granulation

Leaves: odd-pinnately compound with 11-21 leaflets

Flowers: showy and pink, white or purple petals, tightly arranged in a compact raceme with 20-50 flowers per head

Flowering times: June - August

BIRD'S-FOOT TREFOIL

Family: Fabaceae

Common Name: Bird's-Foot Trefoil

Latin Name: *Lotus corniculatus*

Description: A perennial herbaceous plant

Pollen: Light brown 15 μmx20 μm

Nectar: Not a significant honey crop in the UK. Honey is light, with a rapid granulation and tastes like clover.

Other Names: Granny's Toenails, Butter and Eggs, Eggs and Bacon

Leaves: Five leaflets are present but with the central three held conspicuously above the others, hence the use of the name 'trefoil'

Flowers: yellow, 10-16mm long. Five petals; the lateral two are the 'wings', the lower two unite to form the 'keel'. The overall shape of corolla being butterfly-like

Flowering times: June – September

June - Sept

Borage

Family: Boraginaceae

Common Name: Borage

Latin Name: *Borago officinalis*

Description: An annual large, stalked herb

Pollen: White 30μm diameter

Nectar: Not considered a major honey crop. The colour is very pale, with a yellow-grey tint.

Other Names: Starflower

Leaves: It has hairy, ovate leaves that are alternate, simple and 5-15cm long. The leaves are edible and the plant is grown in gardens for that purpose.

Flowers: They are mostly blue, although pink flowers are occasionally observed. The flowers are perfect and complete with five pointed petals that give it a star-like shape.

Flowering Times: June – September

Catmint

Family: Lamiaceae

Common Name: Catmint

Latin Name: *Nepeta spp.*

Description: 250 mostly herbaceous perennials

Pollen: White, 40μm diameter

Nectar: No honey crop

Other Names: Catnip

Leaves: Opposite, heart-shaped, green to grey-green leaves, usually aromatic.

Flowers: The tubular flowers can be lavender, blue, white, pink or lilac, and spotted with tiny lavender-purple dots.

Flowering Times: June – September

Common Mallow

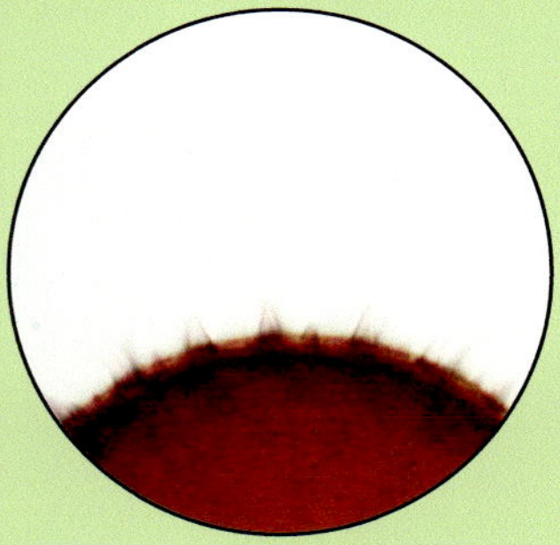

Family: Malvaceae

Common Name: Common Mallow

Latin Name: *Malva sylvestris*

Description: Showy flowers of bright mauve-purple

Pollen: White-grey 100μm diameter

Nectar: No honey crop

Leaves: Borne upon the stem, roundish and three to nine shallow lobes, each 2 – 4cm long.

Flowers: Appear in axillary clusters of 2 – 4 and form irregularly and elongated along the main stem with the flowers at the base opening first. Petals are wrinkly to veined on the backs, more than 20mm long or 15 – 25mm long and 1cm wide

Flowering Times: June – September

June - Sept

Musk Mallow

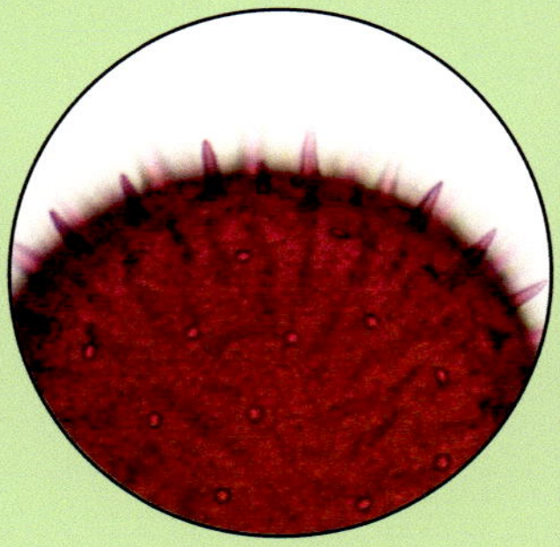

Family: Malvaceae

Common Name: Musk Mallow

Latin Name: *Malva moscata*

Description: Herbaceous perennial plant up to 1m tall

Pollen: White 90 μm diameter

Nectar: No honey crop

Leaves: Alternate 2 – 8cm long and 2 – 8cm broad, palmately lobed with 5 to 7 lobes.

Flowers: Produced in clusters in the leaf axils, each flower 3 – 5cm diameter, with five bright pink petals. They have a distinctive musky odour.

Flowering Times: June – September

Tree Mallow

Family: Malvaceae

Common Name: Tree Mallow

Latin Name: *Lavatera*

Description: Genus with 25 species of flowering plants

Pollen: White 100μm diameter

Nectar: No honey crop

Other Names: Rose Mallow, Royal Mallow, Annual Mallow

Leaves: Spirally arranged and palmately lobed

Flowers: 4 – 12cm diameter with five white, pink or red petals. They are produced in terminal clusters.

Flowering Times: June - September

Mint

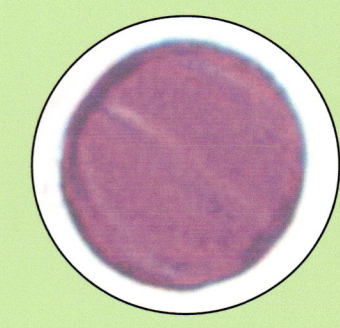

Family: Lamiaceae

Common Name: Mint

Latin Name: *Mentha spp.*

Description: Aromatic, mostly perennial herbs

Pollen: Light brown, 30μm diameter

Nectar: No honey crop

Leaves: The leaves are arranged in opposite pairs, from oblong to lanceolate, often downy and with a serrated margin. Leaf colors range from dark green and grey-green to purple, blue and sometimes pale yellow.

Flowers: The flowers are white to purple and produced in false whorls called verticillasters.

Flowering Times: June – September

Purple Loosestrife

Family: Lythraceae

Common Name: Purple Loosestrife

Latin Name: *Lythrum salicaria*

Description: Robust herbaceous perennial up 1.2m tall

Pollen: Yellow-green 20μm diameter

Nectar: Not a significant crop. Light to dark colour with an aromatic, sharp flavour.

Other Names: Black Blood, Rebel Weed

Leaves: Simple leaves in opposite pairs

Flowers: Star-shaped flowers in leafy racemes. Small, vivid purplish-pink flowers 2cm wide in dense terminal spikes over a long period in summer.

Flowering Times: June – September

Raspberry

Family: Rosaceae

Common Name: Raspberry

Latin Name: *Rubus idaeus*

Description: Generally perennials – biennial stems

Pollen: Light green 25µm diameter

Nectar: A minor honey crop. Flavour and aroma is mild

Leaves: Large pinnately compound leaves with five or seven leaflets. In its second year a stem does not grow taller, but produces several side shoots, which bear smaller leaves with three or five leaflets.

Flowers: Short racemes on the tips of these side shoots, each flower about 1cm diameter with five white petals.

Flowering times: June – September

WILD CARROT

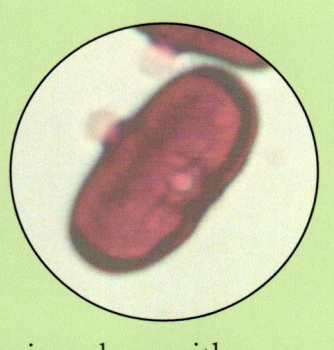

Family: Apiaceae

Common Name: Wild Carrot

Latin Name: *Daucus carota*

Description: A herbaceous, biennial plant

Pollen: Yellow, 10x30μm oblong

Nectar: Not a significant honey crop in the UK. Dark yellow in colour with a distinctive flavour and fragrant aroma like the plant.

Other Names: Bird's Nest, Bishop's Lace, Queen Anne's Lace

Leaves: The leaves are tripinnate, finely divided, lacy and triangular in shape.

Flowers: The flowers are small and creamy white, clustered in flat, dense umbels. The umbels are terminal and 8-10cm wide. They may be pink in bud and may have a reddish or purple flower in the centre of the umbel.

Flowering Times: June - September

Cross-Leaved Heath

Family: Ericaceae

Common Name: Cross-Leaved Heath

Latin Name: *Erica tetralix*

Description: Perennial sub-shrub

Pollen: White, 40μm diameter

Nectar: No honey crop

Other Names: Heath

Leaves: The linear leaves are usually glandular and in whorls of four

Flowers: It has small, pink, drooping, bell-shaped flowers borne in compact clusters at the ends of its shoots.

Flowering Times: June – October

Himalayan Balsam

Family: Balsaminaceae

Common Name: Himalayan Balsam

Latin Name: *Impatiens glandulifera*

Description: Annual plant native to the Himalayas

Pollen: White and oval-shaped, 20 x 10μm in size

Nectar: Not a significant crop – light in colour, mild in flavour and very watery.

Other Names: Policeman's Helmet, Bobby Tops, Copper Tops, Gnome's Hatstand.

Leaves: It has lanceolate leaves that grow 5 to 23cm long. The crushed foliage has a strong, musty smell.

Flowers: Usually the flowers are pink, though some variants are white, with a hooded shape that inspired the nickname 'Policeman's Helmet'. The flowers are usually 3 or 4cm tall and 2cm broad.

Flowering Times: June - October

SWEET ALISON

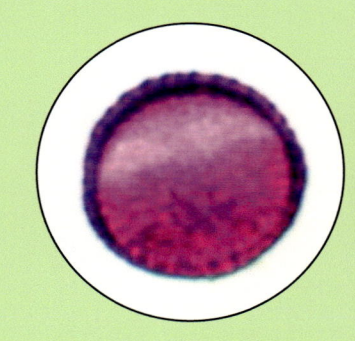

Family: Brassicaceae

Common Name: Sweet Alison

Latin Name: *Lobularia maritima*

Description: Annual plant or short-lived perennial

Pollen: Yellow with a 15μm diameter

Nectar: No honey crop

Other Names: Sweet Alyssum, Alyssum

Leaves: 1 – 4mm long and 3 – 5mm broad. They are alternate, sessile, quite hairy and oval to lanceolate, with an entire margin.

Flowers: About 5mm in diameter, these sweet-smelling flowers have an aroma similar to honey and are usually made up of four rounded white petals and four sepals. However, the petals may also be pink, rose-red, violet or lilac.

Flowering Times: June – October

Dahlia

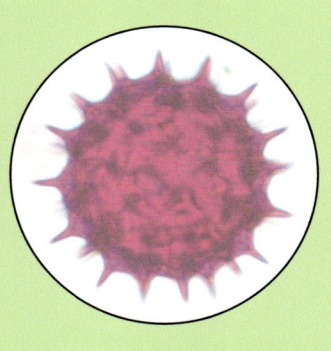

Family: Asteraceae

Common Name: Dahlia

Latin Name: *Dahlia spp.*

Description: 42 species of tuberous and herbaceous, perennials

Pollen: Orange, 40μm diameter

Nectar: No honey crop

Leaves: The stems are leafy, ranging in height from a low as 30cm to more than 1.8-2.4m

Flowers: The flower head is actually a composite with both central disc florets and surrounding ray florets. Each floret is a flower in its own right. Generally, bees prefer the simpler petals as they can reach the nectar more easily.

Flowering Times: June - December

Garden Angelica

Family: Apiaceae

Common Name: Angelica

Latin Name: *Angelica archangelica*

Description: A biennial plant

Pollen: White, 30x10μm oblong

Nectar: Not a significant honey crop in the UK. Dark reddish in colour

Other Names: Garden Angelica, Wild Celery, Norwegian Angelica

Leaves: Comprises of numerous small leaflets divided into three principal groups, each of which is again subdivided into three lesser groups

Flowers: The flowers are small and numerous, grouped into large, globular umbels. They're generally white, but sometimes tinted with yellow or green.

Flowering Times: July

Rudbeckia

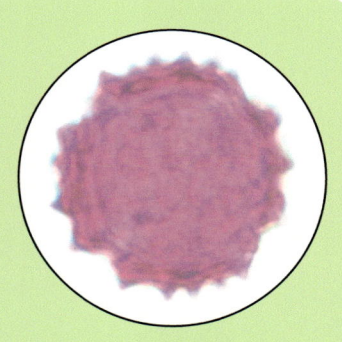

Family: Asteraceae

Common Name: Rudbeckia

Latin Name: *Rudbeckia fulgida*

Description: Herbaceous, mostly perennial plants

Pollen: Orange, 30µm diameter

Nectar: No honey crop

Other Names: Coneflowers, Black-Eyed Susans

Leaves: Leaves are dark green, sparsely but roughly-haired, simple, with sparsely serrated margins

Flowers: On tall stems, this plant has large daisy-like flower-heads with yellow or orange rays surrounding a prominent conical disk. Petals usually point down slightly, away from the centre

Flowering Times: July

YARROW

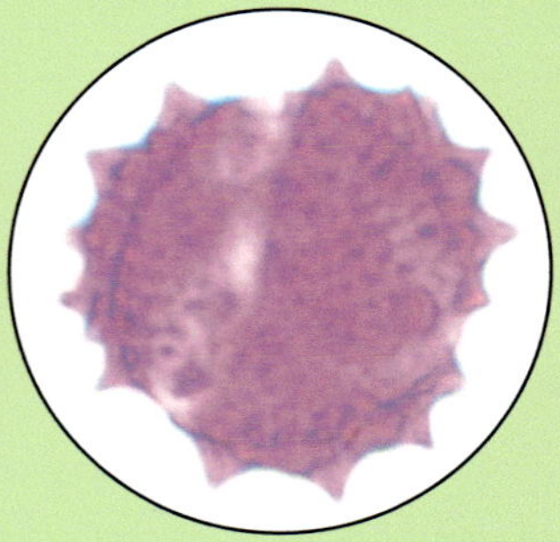

Family: Asteraceae

Common Name: Yarrow

Latin Name: *Achillea millefolium*

Description: An erect, herbaceous perennial

Pollen: Orange, 30μm diameter

Nectar: No honey crop

Other Names: Common Yarrow

Leaves: The leaves are bipinnate or tripinnate, usually 5-20cm long, almost leathery to touch and arranged spirally on the stems.

Flowers: The inflorescence has four to nine phyllaries and contains ray and disk flowers, which are white to pink. The ray flowers are usually ovate to round and found in groups of three to eight.

Flowering Times: July

Bell Heather

Family: Ericaceae

Common Name: Bell Heather

Latin Name: *Erica cinerea*

Description: Low, spreading shrub

Pollen: White-grey, 40μm diameter

Nectar: Major honey source. The honey is port wine in colour with a distinctive pronounced flavour and rapid granulation.

Leaves: Fine needle-like leaves, 4 – 8cm long, arranged in whorls of three.

Flowers: The flowers are bell-shaped, reaching 4 – 7mm long. They are usually fuchsia to magenta in colour but some rare varieties are white.

Flowering Times: July – August

Chinese Bee Tree

Family: Rutaceae

Common Name: Chinese Bee Tree

Latin Name: *Tetradium daniellii*

Description: A large deciduous tree up to 30m tall

Pollen: Yellow, 30μm diameter

Nectar: No honey crop

Other Names: Euodia, Korean Evodia, Bee-Bee Tree

Leaves: Glossy, dark-green, pinnate leaves, 40cm long or more, each with up to eleven elliptic, oval or lance-shaped leaflets. They turn yellow in the autumn.

Flowers: Produced in domed, terminal corymbs to 15cm across, the flowers are white with yellow anthers.

Flowering Times: July – August

CHIVES

Family: Amaryllidaceae

Common Name: Chives

Latin Name: *Allium schoenoprasum* **Description:** Bulb-forming, herbaceous perennial **Pollen:** Yellow, 30x10μm

Nectar: No honey crop

Leaves: It has grass-like, aromatic leaves, which are shorter than the scapes and are also hollow and tubular.

Flowers: There are generally a pale purple and star-shaped with six petals and produced in a dense spherical inflorescence of 10-30 flowers.

Flowering Times: July – August

CLEMATIS

Family: Ranunculaceae

Common Name: Clematis

Latin Name: *Clematis vitalba*

Description: Climbing shrub with branched stems

Pollen: Yellow 20 μm diameter

Nectar: No honey crop

Other Names: Traveller's Joy or Old Man's Beard

Leaves: Pinnately compound, with three to five leaflets which are elliptical in shape with rough-toothed margins.

Flowers: White in colour, the various stamens are clearly visible and the flowers are in clusters. The flowers are around 2cm in width.

Flowering times: July – August

ROSEBAY WILLOWHERB

Family: Onagraceae

Common Name: Rosebay Willowherb

Latin Name: *Chamerion angustifolium*

Description: A perennial herbaceous plant

Pollen: Green 70μm diameter

Nectar: Medium honey crop. Very pale with fine granulation. Flavour mild and very sweet.

Other Names: Fireweed, Epilobium

Leaves: Spirally arranged, entire, narrowly lanceolate and pinnately veined

Flowers: 2-3cm in diameter, slightly asymmetrical, with four magenta to pink petals and four narrower pink sepals behind.

Flowering Times: July – August

Teasel

Family: Caprifoliaceae

Common Name: Teasel

Latin Name: *Dipsacus fullonum*

Description: Flowering plant and cultivar

Pollen: White 75μm diameter

Nectar: No honey crop

Other Names: Wild Teasel, Fuller's Teasel

Leaves: Lanceolate, 20-40cm long and 3-6cm wide with a row of small lines on the underside of the midrib.

Flowers: The inflorescence is a cylindrical array of lavender flowers which dries to a cone of spine-tipped hard bracts. It may be 10cm long.

Flowering Times: July – August

Checkerbloom

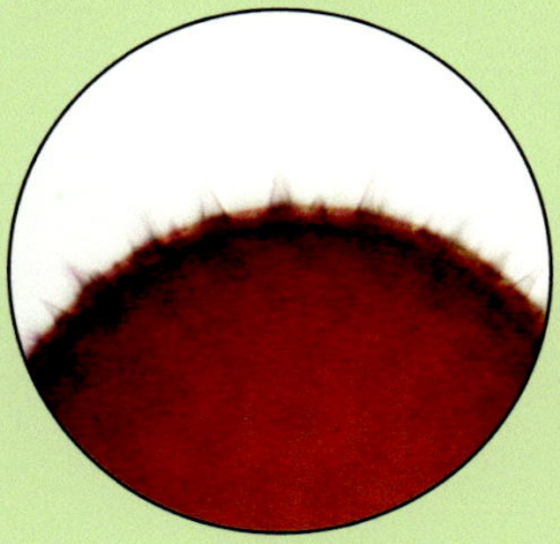

Family: Malvaceae

Common Name: Checkerbloom

Latin Name: *Sidalcea spp.*

Description: Annual or perennial flowering plants

Pollen: Light brown 100µm diameter

Nectar: No honey crop

Other Names: Greek Mallow, Checkermallows, Prairie Mallows

Leaves: Clumps of toothed basal leaves

Flowers: This plant produces erect flowering stems, with five-petalled mallow-type flowers in terminal racemes, in shades of pink, white and purple.

Flowering Times: July - September

Golden Rod

Family: Asteraceae

Common Name: Golden Rod

Latin Name: *Solidago spp.*

Description: Mostly herbaceous perennials found in open areas, up to 120 species

Pollen: Orange, 25µm diameter

Nectar: No honey crop

Leaves: Leaves can differ from species to species. The leaf margins are most commonly entire, but some species may have heavy serration. In some species, the basal leaves are shed before flowering.

Flowers: The flower heads are usually of the radiate type but sometimes discoid. Floret corollas are usually yellow but, in some species, there is white in the ray florets.

Flowering Times: July - September

Hyssop

Family: Lamiaceae

Common Name: Hyssop

Latin Name: *Hyssopus officinalis*

Description: A compact, spreading, semi-evergreen sub-shrub

Pollen: Light-green, 40μm diameter

Nectar: Not a significant crop. Aromatic flavour

Leaves: The leaves are lanceolate, dark green in colour and from 2-2.5cm long

Flowers: The flowers come in bunches of pink, blue, or, more rarely, white fragrant flowers

Flowering Times: July – September

KNAPWEED

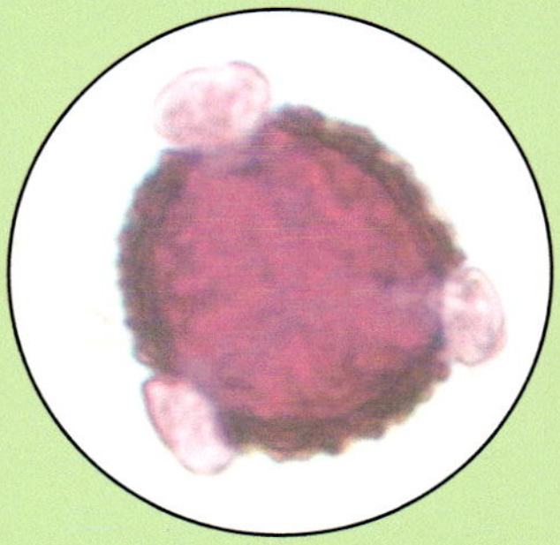

Family: Asteraceae

Common Name: Knapweed

Latin Name: *Centaurea nigra*

Description: A perennial herb

Pollen: White-grey, 50μm diameter

Nectar: No honey crop

Other Names: Lesser Knapweed, Common Knapweed, Black Knapweed, Hardheads

Leaves: The leaves are up to 25cm long, usually deeply-lobed and hairy.

Flowers: The inflorescence contains a few flower heads, each a hemisphere of black or brown bristly phyllaries. The head bears many small, bright purple flowers.

Flowering Times: July – September

Sunflower

Family: Asteraceae

Common Name: Sunflower

Latin Name: *Helianthus spp.*

Description: Usually tall annual or perennial plants that in some species can grow to a height of 3m or more

Pollen: Yellow-orange, 35μm diameter

Nectar: Not a significant honey crop in the UK. Dark yellow in colour with rapid, fine, soft granulation. Flavour mild but distinctive, rather like butter, with a fairly strong aroma.

Leaves: The petiolate leaves are dentate and often sticky. The lower leaves are opposite, ovate or often heart-shaped.

Flowers: Each plant bears one or more wide, terminal capitula (flower heads), with bright yellow ray florets at the outside and yellow or maroon disc florets inside.

Flowering Times: July – September

Tansy

Family: Asteraceae

Common Name: Tansy

Latin Name: *Tanacetum vulgare*

Description: A perennial, herbaceous plant

Pollen: Orange, 30μm diameter

Nectar: No honey crop

Other Names: Common Tansy, Bitter Buttons, Cow Bitter, Golden Buttons

Leaves: The leaves are alternate, 10-15cm long and are pinnately lobed, divided almost to the centre into about the seven pairs of segments.

Flowers: The roundish, flat-topped, button-like, yellow flowering heads are produced in terminal clusters

Flowering Times: July – September

Ragwort

Family: Asteraceae

Common Name: Ragwort

Latin Name: *Senecio jacobaea*

Description: A weed of paddocks and gardens

Pollen: Yellow, 25μm diameter

Nectar: No honey crop

Other Names: Common Ragwort

Leaves: It has finely divided leaves with a basal rosette of d eeply-cut, toothed leaves.

Flowers: Large, flat-topped clusters of yellow daisy-like flowers

Flowering Times: July – October

Sea Aster

Family: Asteraceae

Common Name: Sea Aster

Latin Name: *Tripolium pannonicum*

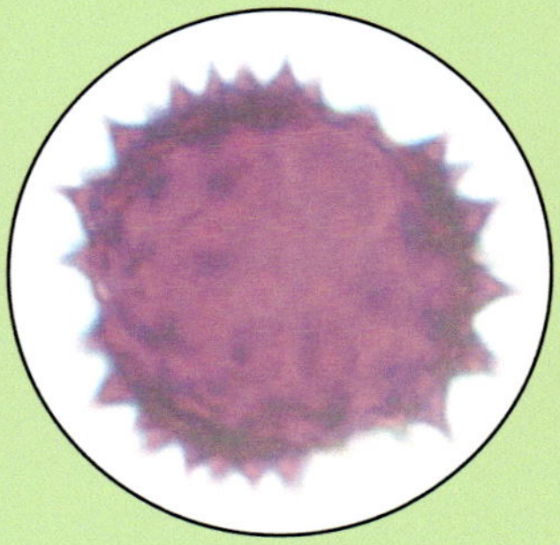

Description: A short-lived perennial herb

Pollen: Yellow-orange 50μm diameter

Nectar: No honey crop

Other Names: Seashore Aster, Aster Tripolium, Aster Pannonicus

Leaves: Growing an alternate pattern, the lowest leaves are stalked but upper leaves are stalkless.

Flowers: A single flower-like 2-3cm capitula surrounded by involucral bracts. The capitula's ray-florets are pink or blue, occasionally white.

Flowering Times: July – October

SEA LAVENDER

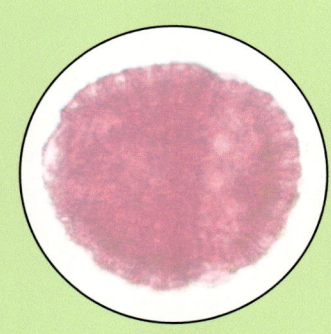

Family: Plumbaginaceae

Common Name: Sea Lavender

Latin Name: *Limonium vulgare*

Description: Herbaceous, perennial plants

Pollen: Light yellow, 50μm diameter

Nectar: No honey crop

Other Names: Common Sea-Lavender, Sea Thrift

Leaves: The leaves are simple, entire to lobed, and from 1 – 30cm long and 0.5 – 10cm broad.

Flowers: Produced on a branched panicle or corymb, the individual flowers are small (4 – 10mm long). The flower colour is pink or violet to purple in most species.

Flowering Times: July - October

July - Oct

Verbena

Family: Verbenaceae

Common Name: Verbena

Latin Name: *Verbena bonariensis*

Descri`ption: A Flowering herbaceous perennial

Pollen: Light brown 40μm diameter

Nectar: No honey crop

Other Names: Purpletop, South American Vervain

Leaves: Ovate to ovate-lanceolate with a toothed margin and grow up to 10cm long.

Flowers: Fragrant lavender to rose-purple flowers in tight clusters located on terminal and axillary stems.

Flowering Times: July – November

Bilberry

Family: Ericaceae

Common Name: Bilberry

Latin Name: *Vaccinium myrtillus*

Description: A species of shrub with small, edible, dark berries

Pollen: White, 40μm diameter

Nectar: No honey crop

Other Names: Wimbleberry, Whortleberry, European Blueberry, Blue Whortleberry

Leaves: The leaves are finely-toothed and prominently veined on the lower surface

Flowers: Borne singly in leaf axils on 2-3mm long pedicels. The corolla is pink and shaped like an urn.

Flowering Times: August – September

Ling Heather

Family: Ericaceae

Common Name: Ling Heather

Latin Name: *Calluna vulgaris*

Description: Low-growing perennial shrub

Pollen: Grey-brown, 40μm diameter

Nectar: Major honey source. Reddish amber colour with slow coarse granulation. Slightly bitter flavour with a strong aroma from floral to medicinal.

Other Names: Common Heather, Ling, Heather

Leaves: Small scale-leaves (less than 2 – 3cm long) borne in opposite and decussate pairs

Flowers: Normally mauve but white-flowered plants also occur occasionally. They are terminal in racemes with sepal-like bracts at the base.

Flowering Times: August – October

STRAWBERRY TREE

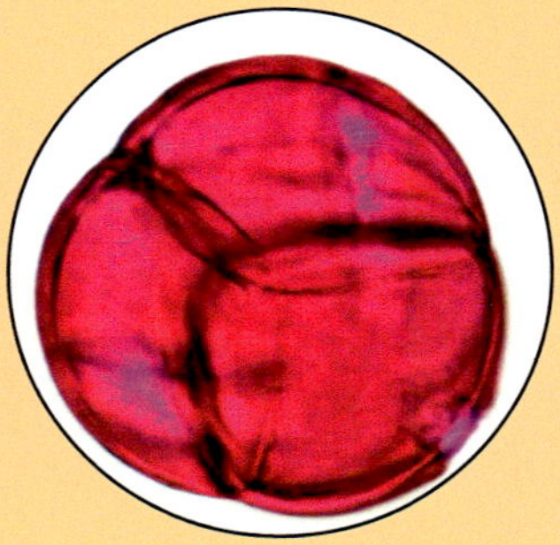

Family: Ericaceae

Common Name: Strawberry Tree

Latin Name: *Arbutus unedo*

Description: Evergreen shrub or small tree

Pollen: Brown, 50μm diameter

Nectar: No honey crop

Other Names: Cain or Cane Apple, Killarney Strawberry Tree

Leaves: They are dark green and glossy, 5 – 10cm long and 2 – 3cm broad, with a serrated margin.

Flowers: Usually white, though occasionally pale pink and bell-shaped, the flowers are produced in panicles of 10 to 30, each growing 4 – 6mm in diameter.

Flowering Times: September – November

Ivy

Family: Araliaceae

Common Name: Ivy

Latin Name: *Hedera helix*

Description: A clinging evergreen vine

Pollen: Orange, 35μm diameter

Nectar: No significant honey crop. The colour is light; both the flavour and the fragrance are repellent. It granulates quickly and very hard.

Other Names: Common Ivy, English Ivy, European Ivy

Leaves: Leaves are alternate, 50-100mm long, with a 15-20mm petiole (stalk)

Flowers: Individually small and greenish-yellow, they grow in umbels that range from 3-5cm in diameter.

Flowering Times: September - November

Mahonia

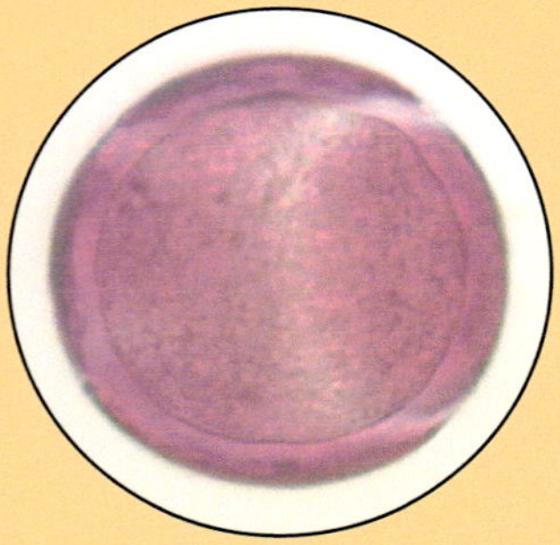

Family: Beriberidaceae

Common Name: Mahonia

Latin Name: *Mahonia spp.*

Description: Genus of about 70 species of shrubs

Pollen: Green, 35µm diameter

Nectar: Not a significant honey crop in the UK. Light amber in colour.

Leaves: Typically they have large, pinnate leaves 10-50cm long with five to fifteen leaflets.

Flowers: The flowers are produced in racemes, which are 5-20cm long.

Flowering Times: November - February

Winter

Winter Heath

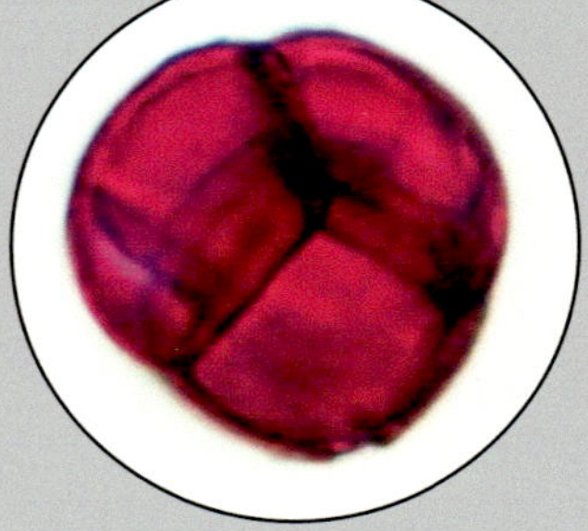

Family: Ericaceae

Common Name: Winter Heath

Latin Name: *Erica carnea*

Description: Low-growing, spreading sub-shrub

Pollen: White-grey, 30μm diameter

Nectar: Not a significant crop. Light to dark yellow colour, strong in favour and aroma.

Other Names: Winter Flowering Heather, Winter Heath, Spring Heath, Alpine Heath

Leaves: Borne in whorls of four, the evergreen, 4 – 8mm leaves of this plant are thin and flat, giving them a needle-like appearance.

Flowers: The individual flower is a slender bell shape that grows to 4 – 6mm long and usually dark pink, though can occasionally be white. They are produced in racemes.

Flowering Times: December – March

Darley Dale Heath

Family: Ericaceae

Common Name: Darley Dale Heath

Latin Name: *Erica darleyensis*

Description: Low-growing, spreading sub-shrub

Pollen: Yellow, 30μm diameter

Nectar: No honey crop

Other Names: Heath

Leaves: The small, lance-shaped leaves are mid-green with pink tips in spring.

Flowers: This plant grows clusters of lilac-pink, urn-shaped flowers.

Flowering Times: December – May

SNOWDROP

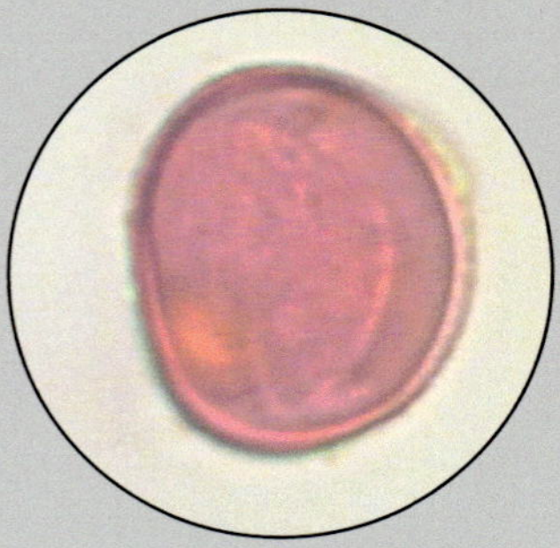

Family: Amaryllidaceae

Common Name: Snowdrop

Latin Name: *Galanthus nivalis*

Description: Perennial, herbaceous bulbous plants

Pollen: Orange, 30x20μm

Nectar: No honey crop

Other Names: Common Snowdrop

Leaves: Each plant produces two linear, or very narrowly lanceolate, greyish-green leaves

Flowers: A solitary, pendulous, bell-shaped, white flower, held on a slender pedicel.

Flowering Times: January – March

Gorse

Family: Fabaceae

Common Name: Gorse

Latin Name: *Ulex europaeus*

Description: Species of thorny evergreen shrubs

Pollen: Orange-brown 40 µm

Nectar: No honey crop

Other Names: Furze, Whin

Leaves: The leaves of young plants are trifoliate, but in mature plants they reduced to scales or small spines

Flowers: Yellow flowers, generally showy, some with a very long flowering season. Gorse flowers have a distinctive coconut scent

Flowering times: January – June

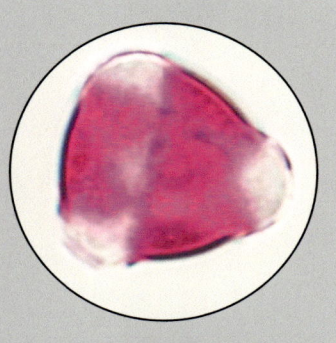

CROCUS

Family: Iridaceae

Common Name: Crocus

Latin Name: *Crocus spp.*

Description: 90 species of dwarf, deciduous perennials

Pollen: Orange, 70-100μm diameter

Nectar: No honey crop

Leaves: The grass-like, ensiform leaf shows generally a white central stripe along the leaf axis.

Flowers: The cup-shaped, solitary, slaverform flower tapers off into a narrow tube. Their colours vary enormously, although lilac, mauve, yellow, and white are predominant.

Flowering Times: February - March

Musk Willow

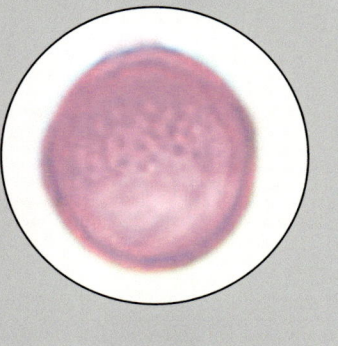

Family: Salicaceae

Common Name: Willow

Latin Name: *Salix aegyptiaca*

Description: Deciduous shrubs and trees

Pollen: Yellow 20μm diameter

Nectar: No honey crop

Other Names: Musk Willow, Calaf of Persia Willow

Leaves: Simple, shiny and alternate. 5-15cm long and 3-6cm wide

Flowers: Dioecious. Catkins are fragrant and grey. Males grow to 30mm long with yellow anthers; females to 75mm long

Flowering Times: February – March

WINTER ACONITE

Family: Ranunculaceae

Common Name: Winter Aconite

Latin Name: *Eranthis hyemalis*

Description: Small tuberous perennials

Pollen: Yellow 40 μm diameter

Nectar: No honey crop

Other names: Winter hellebore, Winterling

Leaves: Palmately or Pinnately lobed basal leaves and cup-shaped flowers held above a collar of deeply lobed stem leaves

Flowers: Large 2-3cm, yellow, cup-shaped flowers held above a collar of 3 leaf-like bracts.

Flowering times: February – March

Blackthorn

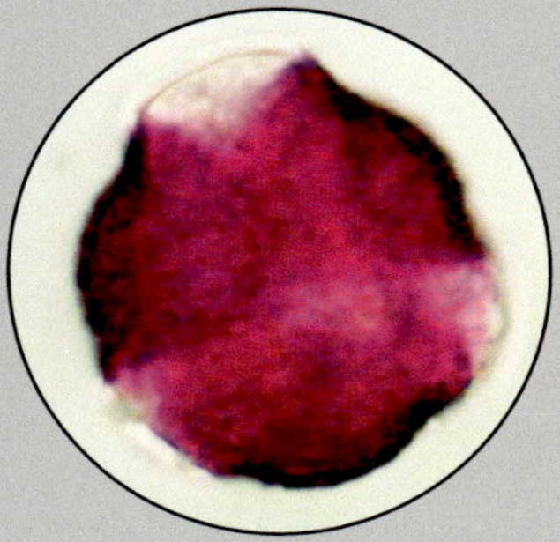

Family: Rosaceae

Common Name: Blackthorn

Latin Name: *Prunus spinosa*

Description: Spiny, densely branched trees that grow up to 7m tall

Pollen: Orange 40μm diameter

Nectar: No honey crop

Other Names: Sloe

Leaves: Slightly wrinkled, oval, toothed pointed at the tip and tapered at the base.

Flowers: White flowers appear on short stalks before the leaves in March and April, either singularly or in pairs.

Flowering times: February – April

Feb - Apr

LENTEN ROSE

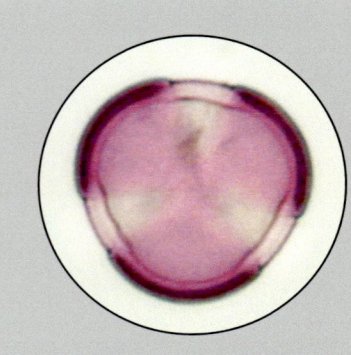

Family: Ranunculaceae

Common Name: Lenten Rose

Latin Name: *Helleborus orientalis*

Description: Perennial elegant garden plants

Pollen: Yellow 30 μm diameter

Nectar: No honey crop

Other Names: Christmas Rose

Leaves: They have no true leaves on their flower stalks, although there are leafy bracts where the flower stalks branch.

Flowers: Pretty, pendent or outward-facing, saucer-shaped flowers

Flowering times: February – April

Pieris

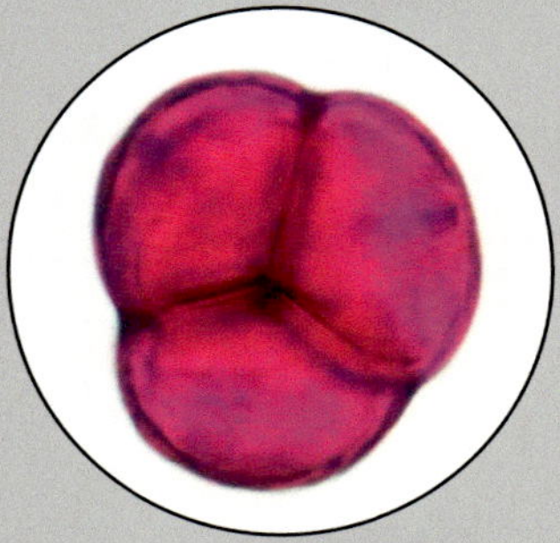

Family: Ericaceae

Common Name: Pieris

Latin Name: *Pieris spp.*

Description: Broad-leaved, evergreen shrub

Pollen: White, 30μm diameter

Nectar: No honey crop

Other Names: Andromedas, Fetterbushes

Leaves: The leaves are spirally arranged, often appearing to be in whorls at the end of each shoot. They are lanceolate-ovate, 2 – 10cm long and 1 – 3.5cm broad. In some species, the leaves at the top are red, changing to green at the age.

Flowers: The flowers are white or pink, bell-shaped, 5 – 15mm long and arranged in racemes 5 – 12cm long.

Flowering Times: February – April

Plant Index

Apple, *Malus pumila* 30

Asparagus, *Asparagus officinalis* 42

Aubretia, *Aubretia deltoidea* 26

Barberry, *Berberis spp.* 31

Bell Heather, *Erica cinerea* 94

Bilberry, *Vaccinium myrtillus* 111

Bird's-Foot Trefoil, *Lotus corniculatus* 76

Blackberry, *Rubus fruticosus* 60

Blackcurrant, *Ribes nigra* 43

Blackthorn, *Prunus spinosa* 128

Bluebell, *Hyacinthoides non-scripta* 32

Blueberry, *Vaccinium corymbosum* 44

Borage, *Borago officinalis* 77

Broad Bean, *Vicia faba* 45

Broom, *Cytisus scoparius* 33

Calico Bush, *Kalmia latifolia* 56

Catmint, *Nepeta spp.* 78

Charlock, *Sinapis Arvensis* 57

Checkerblooms, *Sidalcea Spp.* 100

Cherry Laurel, *Prunus laurocerasus* 34

Cherry, *Prunus spp.* 22

Chinese Bee Tree, *Tetradium daniellii* 95

Plant Index A - C

Chives, *Alluim schoenoprasum* 96

Clematis, *Clematis vitalba* 97

Comfrey, *Symphytum spp.* 53

Common Lungwort, *Pulmonaria officinalis* 24

Common Mallow, *Malva sylvestris* 79

Cotoneaster, *Cotoneaster spp.* 46

Cranberry, *Vaccinium oxycoccos* 65

Crocus, *Crocus spp.* 125

Cross-Leaved Heath, *Erica tetralix* 86

Dandelion, *Taraxacum officinale* 28

Darley Dale Heath, *Erica darleyensis* 120

Dhalia, *Dhalia spp.* 89

Forget-Me-Not, *Myosotis spp.* 40

Garden Angelica, *Angelica archangelica* 91

Golden Rod, *Solidago spp.* 101

Gooseberry, *Ribes uva-crispa* 23

Gorse, *Ulex europaeus* 123

Hawthorn, *Crataegus spp.* 27

Himalayan Balsam, *Impatiens glandulifera* 87

Holly, *Ilex aquifolium* 47

Hollyhock, *Alcea rosea* 70

Horse Chestnut, *Aesculus hippocastanum* 48

Hyssop, *Hyssopus officinalis* 102

Ivy, *Hedera helix* 115

Kanpweed, *Centaurea nigra* 103

Lavender, *Lavandula angustifolia* 71

Lemon Balm, *Melissa officinalis* 72

Lenten Rose, *Helleborus orientalis* 129

Lime, *Tilia spp.* 66

Ling Heather, *Calluna vulgaris* 112

Lovage, *Levisticum officinale* 54

Mahonia, *Mahonia spp.* 117

Maple, *Acer spp.* 49

Marjoram, *Origanum majorana* 73

Mexican Orange Blossom, *Choisya ternata* 50

Mint, *Mentha spp.* 82

Musk Mallow, *Malva moscata* 80

Musk Willow, *Salix aegyptiaca* 126

Oilseed Rape, *Brassica napus* 35

Pear, *Pyrus spp.* 21

Phacelia, *Phacelia tanacetifolia* 61

Phlomis, *Phlomis* 62

Pieris, *Pieris spp.* 130

Poppy, *Papaver spp.* 58

Privet, *Ligustrum vulgare* 74

Purple Loosestrife, *Lythrum salicaria* 83

Pussy Willow, *Salix caprea* 25

Ragwort, *Senecio jacobaea* 106

Raspberry, *Rubus idaeus* 84

Red Clover, *Trifolium Ppatense* 68

Redcurrant, *Ribes ruba* 36

Rhododendron, *Rhododendron ponticum* 51

Rosebay Willowherb, *Chamerion angustifolium* 98

Rosemary, *Rosmarinus officinalis* 37

Rowan, *Sorbus spp.* 52

Rudbeckia, *Rudbeckia fulgida* 92

Sage, *Salvia officinalis* 55

Sainfoin, *Onobrychis viciifolia* 75

Sea Aster, *Tripolium pannonicum* 107

Sea Lavender, *Limonium vulgare* 108

Self-Heal, *Prunella grandiflora* 67

Snowdrop, *Galanthus nivalis* 122

Strawberry Tree, *Arbutus unedo* 114

Strawberry, *Fragaria spp.* 39

Sun lower, *Helianthus spp.* 104

Sweet Alison, *Lobularia maritima* 88

Sycamore, *Acer pseudoplatanus* 38

Tansy, *Tanacetum vulgare* 105

Teasel, *Dipsacus fullonum* 99

Thyme, *Thymus spp.* 59

Tree Mallow, *Lavatera* 81

Verbena, *Verbena bonariensis* 10

Vipers Bugloss, *Echium vulgare* 63

White Clover, *Trifolium Rrpens* 69

Wild Carrot, *Daucus carota* 85

Winter Aconite, *Eranthis hyemalis* 127

Winter Heath, *Erica carnea* 119

Yarrow, *Achillea millefolium* 93